Jessica Richard

Toxicity of micropollutants during advanced oxidation processes

AF061338

Jessica Richard

Toxicity of micropollutants during advanced oxidation processes

Südwestdeutscher Verlag für Hochschulschriften

Impressum / Imprint
Bibliografische Information der Deutschen Nationalbibliothek: Die Deutsche Nationalbibliothek verzeichnet diese Publikation in der Deutschen Nationalbibliografie; detaillierte bibliografische Daten sind im Internet über http://dnb.d-nb.de abrufbar.
Alle in diesem Buch genannten Marken und Produktnamen unterliegen warenzeichen-, marken- oder patentrechtlichem Schutz bzw. sind Warenzeichen oder eingetragene Warenzeichen der jeweiligen Inhaber. Die Wiedergabe von Marken, Produktnamen, Gebrauchsnamen, Handelsnamen, Warenbezeichnungen u.s.w. in diesem Werk berechtigt auch ohne besondere Kennzeichnung nicht zu der Annahme, dass solche Namen im Sinne der Warenzeichen- und Markenschutzgesetzgebung als frei zu betrachten wären und daher von jedermann benutzt werden dürften.

Bibliographic information published by the Deutsche Nationalbibliothek: The Deutsche Nationalbibliothek lists this publication in the Deutsche Nationalbibliografie; detailed bibliographic data are available in the Internet at http://dnb.d-nb.de.
Any brand names and product names mentioned in this book are subject to trademark, brand or patent protection and are trademarks or registered trademarks of their respective holders. The use of brand names, product names, common names, trade names, product descriptions etc. even without a particular marking in this works is in no way to be construed to mean that such names may be regarded as unrestricted in respect of trademark and brand protection legislation and could thus be used by anyone.

Coverbild / Cover image: www.ingimage.com

Verlag / Publisher:
Südwestdeutscher Verlag für Hochschulschriften
ist ein Imprint der / is a trademark of
OmniScriptum GmbH & Co. KG
Heinrich-Böcking-Str. 6-8, 66121 Saarbrücken, Deutschland / Germany
Email: info@svh-verlag.de

Herstellung: siehe letzte Seite /
Printed at: see last page
ISBN: 978-3-8381-3799-5

Zugl. / Approved by: Duisburg-Essen, Universität, Diss., 2013

Copyright © 2014 OmniScriptum GmbH & Co. KG
Alle Rechte vorbehalten. / All rights reserved. Saarbrücken 2014

Table of Contents

I List of Figures -- *5*

II List of Tables -- *11*

III List of Abbreviations -- *13*

1.   *Summary* -- *19*
2.   *Introduction* -- *21*
    2.1.   **Micropollutants in waste water treatment plant effluents and surface water** -- 22
        2.1.1. Guidelines for water quality control -------------------------------------- 22
        2.1.2. Emission of micropollutants -- 23
        2.1.3. Substances detected in waste water treatment plant effluents ------ 24
        2.1.4. Waste water quality control --- 24
        2.1.5. Common waste water treatment -- 26
        2.1.6. Advanced waste water treatment ------------------------------------- 28
            2.1.6.1. Filtration -- 28
            2.1.6.2. Sorption: Activated Carbon --------------------------------- 30
            2.1.6.3. Advanced Oxidation Processes (AOP) --------------------- 31
    2.2.   **Toxicological testing for the determination of biological effects of micropollutants** --- 36
        2.2.1. Cytotoxicity --- 44
        2.2.2. Genotoxicity -- 44
        2.2.3. Mutagenicity --- 46
        2.2.4. Estrogenicity --- 47
3.   *Aims of the study* --- *49*
4.   *Material and Methods* --- *50*
    4.1.   **Solutions** -- 50
        4.1.1. Cell Culture -- 50
            4.1.1.1. CHO cells --- 50
            4.1.1.2. T47D cells -- 50
        4.1.2. PAN I Multitox Test -- 51
        4.1.3. MTT Test -- 51
        4.1.4. Alkaline Comet Assay --- 52
        4.1.5. Ames Test --- 53
        4.1.6. ER Calux -- 53
    4.2.   **Cell Culture methods** --- 54
        4.2.1. CHO cells --- 54
            4.2.1.1. Thawing --- 54
            4.2.1.2. Subculturing -- 54
            4.2.1.3. Freezing -- 55
        4.2.2. T47D cells -- 55
            4.2.2.1. Thawing --- 56

  4.2.2.2. Subculturing ----- 56
  4.2.2.3. Freezing ----- 56
 4.3. **Toxicological methods** ----- 57
  4.3.1. Cytotoxicity ----- 57
   4.3.1.1. PAN I: LDHe – XTT – NR – SRB ----- 57
   4.3.1.2. MTT Test ----- 63
  4.3.2. Degree of Cytotoxicity ----- 65
  4.3.3. Genotoxicity ----- 66
   4.3.3.1. Alkaline Comet Assay ----- 66
  4.3.4. Mutagenicity ----- 70
   4.3.4.1. Ames Test ----- 70
  4.3.5. Estrogenicity ----- 73
   4.3.5.1. ER Calux ----- 73

 4.4. **Oxidation of water samples** ----- 76
  4.4.1. Ozonation ----- 78
   4.4.1.1. Laboratory scale ozonation ----- 78
   4.4.1.2. Pilot scale ozonation (IUTA e.V.) ----- 79
  4.4.2. UV/H_2O_2 oxidation ----- 79
   4.4.2.1. Laboratory scale UV oxidation with and without H_2O_2 (IUTA e.V.) -- 79
   4.4.2.2. UV/H_2O_2 oxidation of TPP, TBEP and TCPP (UDE) ----- 79
   4.4.2.3. Pilot scale UV oxidation with and without H_2O_2 (IUTA e.V.) ----- 80
   4.4.2.4. UV and UV/H_2O_2 oxidation using the flow through system (IUTA e.V.) 80
   4.4.2.5. UV and UV/H_2O_2 oxidation at the flow through system (Waste water treatment plant) ----- 80

 4.5. **Extraction methods** ----- 81
  4.5.1. Solid Phase Extraction (SPE) ----- 81
  4.5.2. Liquid-liquid-extraction (LLE) ----- 81

 4.6. **Analytical Chemistry** ----- 82
  4.6.1. HPLC-MS/MS ----- 82
  4.6.2. GC-MS ----- 85

 4.7. **Peroxide testing** ----- 86

 4.8. **Waste water treatment plant effluent used for testing** ----- 87

 4.9. **Sample preparation before toxicological testing** ----- 87

5. ***Results*** ----- 88

 5.1. **Adaptation of toxicological methods** ----- 88
  5.1.1. MTT Test, PAN I and Alkaline Comet Assay ----- 88
  5.1.2. ER Calux ----- 89
  5.1.3. Ames Test ----- 90
  5.1.4. H_2O_2 ----- 90
  5.1.5. Contaminations ----- 92

 5.2. **Matrix controls** ----- 93

- 5.2.1. HPLC-water --- 93
- 5.2.2. Waste water treatment plant effluent --- 95

5.3. ß-Blocker --- 97
- 5.3.1. Atenolol --- 97
- 5.3.2. Metoprolol --- 98

5.4. Estrogenic substances --- 100
- 5.4.1. Ethinylestradiol --- 101
- 5.4.2. Bisphenol A --- 102

5.5. Antibiotics --- 104
- 5.5.1. Sulfamethoxazole --- 105
- 5.5.2. Ciprofloxacin --- 109
- 5.5.3. Ofloxacin --- 114
- 5.5.4. Triclosan and 2,4-Dichlorophenol --- 116

5.6. Biocides --- 118
- 5.6.1. Irgarol 1051 --- 118
- 5.6.2. Terbutryn --- 119

5.7. Musk fragrances --- 120
- 5.7.1. AHTN --- 120
- 5.7.2. HHCB --- 122

5.8. Organophosphates --- 124
- 5.8.1. Tris(2-chloro-1-methylethyl) phosphate (TCPP) --- 124
- 5.8.2. Tris(2-chloroethyl)phosphate (TCEP) --- 125
- 5.8.3. Triphenyl phosphate (TPP) --- 126

6. *Discussion* --- 128

6.1. Endocrine disruption --- 129

6.2. Antibacterial activity during advanced oxidation processes --- 131

6.3. Mixture effects --- 133

6.4. Use of toxicological *in vitro* methods for water quality control --- 135

6.5. Applicability of ozonation and UV/H_2O_2 oxidation during waste water treatment --- 137

6.6. Concentration of water samples --- 139

6.7. ß-blocker --- 140

6.8. Estrogenic substances --- 142
- 6.8.1. Ethinylestradiol --- 142
- 6.8.2. Bisphenol A --- 143

6.9. Antibiotics --- 145

- 6.9.1. Sulfamethoxazole --- 145
- 6.9.2. Fluoroquinolone antibiotics --- 146
- 6.9.3. Triclosan and 2.4-Dichlorophenol --- 149

6.10. Biocides --- 151
- 6.10.1. Irgarol 1051 --- 151
- 6.10.2. Terbutryn --- 153

6.11. Musk fragrances --- 155
- 6.11.1. AHTN and HHCB --- 155

6.12. Organophosphates: TCEP, TCPP and TPP --- 157

6.13. Conclusions and future directions --- 160

7. *References* --- *161*
8. *Annex* --- *185*
   - 8.1. List of Chemicals --- 185
   - 8.2. List of Materials --- 188
   - 8.3. List of Equipment --- 189

Acknowledgements --- *191*

I List of Figures

Figure 1: Routes of micropollutant contamination of surface waters 23

Figure 2: Schematic of a waste water treatment plant. AOP = Advanced Oxidation Processes 27

Figure 3: Membrane separation processes overview [42]. 28

Figure 4: Suitability of water treatment technologies according to the chemical oxygen demand (COD in g/L) [51]. 32

Figure 5: Reaction pathways of the OH° (encircled) formation by ozone [48] 33

Figure 6: Oxidation and reaction pathways of the UV treatment without (a) and with (b) H_2O_2 34

Figure 7: Fenton reaction 35

Figure 8: Photo-Fenton reaction 35

Figure 9: Reaction pathway of the TiO_2 photocatalysis 36

Figure 10: Testing strategy proposed by the Committee on Toxicity Testing and Assessment of Environmental Agents [74] 38

Figure 11: Bioanalytical tools for the assessment of human health risks adapted from Escher et al. 2011 [26] 39

Figure 12: Level of toxicological effect and time until manifestation of effects in regard to the method, adapted from Kase et al. 2009 [71]. 40

Figure 13: Schematic of toxicological and chemical (grey box) testing as part of this project. Chemical analyses have been performed by the working groups of the IUTA e.V. or the University of Duisburg-Essen. 43

Figure 14: CHO cells; passage 30 54

Figure 15: T47D cells; passage 22 56

Figure 16: Chemical equation of the XTT cleavage to XTT formazan by succinate dehydrogenases 58

Figure 17: Structural formula of neutral red 59

Figure 18: Structural formula of Sulforhodamine B 59

Figure 19: Layout of the 96 well plate seeded with CHO cells (20000 cells/0.2 mL). B = 200 µL blank control: culture medium + solvent without cells; GC = 200 µL cell growth control: culture medium + cells; SC = 200 µL solvent control: culture medium + cells + solvent; TL = 200 µL positive control: culture medium + cells + 1 % Triton® X-100 60

Figure 20: Equation used to calculate the extracellular NADH amount 62

Figure 21: Equation used to calculate the percent of viable cells 62

Figure 22: Chemical equation of the MTT cleavage to MTT formazan 64

Figure 23: Comet Assay analysis of undamaged (a) and damaged (b) CHO cells stained with SYBR Green® and analyzed using the Comet Assay 4 Software. 67

Figure 24: a) Gelbond film containing eight microgels. b) Scoring pattern of a single microgel. pos = positive control; neg = negative control; 1 – 6 = gels containing cells exposed to the samples. 68

Figure 25: 384-Well plate after 48 h of exposure. The medium of the positive wells containing bacteria with reversed mutations has changed to yellow and the medium of the wells without mutated bacteria is purple. 72

Figure 26: ER Calux exposure plate. E2 = 17ß-ethinylestradiol; E2-0 to E2-30 = 17ß-ethinylestradiol concentration [pM/well]; DMSO = wells containing 1 % DMSO; medium = wells containing medium only; 1 – 10 = wells containing the cells exposed to different samples. 74

Figure 27: Peroxide Test-strips (Quantofix®). Left: negative for peroxides; right: peroxides detected 86

Figure 28: Luminescence measurement of a Luciferase standard using a white and a black 96-well plate. 90

Figure 29: a) Cytotoxic effects of UV/H_2O_2 treated samples. b) DNA damaging effects of UV/H_2O_2 treated samples. The addition of catalase resulted in the elimination of toxic effects. 91

Figure 30: Blood agar plates with bacterial growth after plating an unfiltered WWTP effluent sample (a) and a plate without colonies after plating a filtered sample (b) incubated for 24 h at 37 °C. 92

Figure 31: No cytotoxicity (a) and genotoxicity (b) of HPLC-water before and after ozonation (2 mg O_3/L) or UV/H_2O_2 treatment. 93

Figure 32: No genotoxicity of HPLC water extracts (C18 and Strata X) before and after 60 mintes of ozonation 94

Figure 33: Waste water treatment plant effluent before and after oxidative treatment. No cytotoxic (a) or genotoxic (b) effects. 95

Figure 34: No genotoxicity of WWTP extracts (C18 and Strata X) of WWTP effluent before and after ozonation 96

Figure 35: No cytotoxicity (a) and genotoxicity (b) of 0.2 mg/L Atenolol in HPLC-water before and after ozonation (2 mg O_3/L). 98

Figure 36: a) No cytotoxicity or genotoxicity (b) of HPLC-water containing 1.4 mg/L Metoprolol before and after ozonation or 0.1 mg/L Metoprolol before and after UV/H_2O_2 treatment for 60 minutes. 99

Figure 37: a) No cytotoxicity or genotoxicity (b) of WWTP effluent with 1.4 mg/L Metoprolol before and after ozonation or of 0.1 mg/L Metoprolol after UV/H_2O_2 treatment for 60 minutes. 99

Figure 38: No cytotoxicity of 1.5 mg/L Ethinylestradiol before and after UV/H_2O_2 treatment or ozonation (1-10 mg/L O_3). 101

Figure 39: a) No cytotoxic effects of HPLC-water containing 1.4 mg/L Bisphenol A after ozonation, but cytotoxic effects after UV/H_2O_2 treatment. b) No genotoxic effects of tested samples. No cytotoxic (c) or genotoxic (d) effects of WWTP effluent containing Bisphenol A before and after ozonation or UV/H_2O_2 treatment.▼ cytotoxic effects ;# = not tested 103

Figure 40: Sulfamethoxazole (0.1 mg/L) in HPLC-water before and after UV/H_2O_2 treatment. No cytotoxicity (a) or genotoxicity (b) detected. 106

Figure 41: Sulfamethoxazole (1.4 mg/L) in HPLC-water before and after ozonation. No cytotoxicity (a) or genotoxicity (b) detected. 107

Figure 42: No cytotoxicity (a) and genotoxicity (b) of 1.4 mg/L Sulfamethoxazole in WWTP effluent before (0 min) and after (60 min) ozonation (5 mg/L O_3). 108

Figure 43: No cytotoxicity (a) and genotoxicity (b) of 0.1 mg/L Sulfamethoxazole in WWTP effluent before and after UV/H_2O_2 treatment. 109

Figure 44: a) Cytotoxic effects of 1.7 mg/L Ciprofloxacin in HPLC-water after 60 min UV/H_2O_2 treatment. b) No genotoxic effects before and after a maximum of 30 min UV/H_2O_2 oxidation. c) No cytotoxic or (d) genotoxic effects of 1.4 mg/L of Ciprofloxacin in HPLC-water before and after ozonation. ▼ Cytotoxic; n.t. = not tested 110

Figure 45: WWTP effluent containing 1.4 mg/L Ciprofloxacin show no cytotoxic (a) and genotoxic (b) effects after ozonation (5 mg/L). 111

Figure 46: a) No cytotoxic and b) genotoxic effects of untreated and UV/H_2O_2 (Hg-LP, 15 W; 1 g/L H_2O_2) treated WWTP effluent with 0.1 mg/L Ciprofloxacin. 112

Figure 47: Cytotoxicity (a) and genotoxicity (b) testing of HPLC-water containing 18 mg/L Ofloxacin before and after ozonation (10 mg/L O_3). No cytotoxicity (c) or genotoxicity (d) of 18 mg/L Ofloxacin in HPLC water before and after UV/H_2O_2 treatment. 114

Figure 48: No cytotoxicity (a) or genotoxicity (b) of 18 mg/L Ofloxacin in WWTP effluent before (Matrix control) and after 60 min of ozonation (10 mg/L O_3) or UV/H_2O_2 treatment. 115

Figure 49: HPLC-MS/MS Q1-Scan chromatogram of a sample with different triclosan:ozone ratios of 1:1, 1:3 and 1:5 [173]. 116

Figure 50: a) No cytotoxicity of 0 – 100 μg/L Triclosan or 2,4-Dichlorophenol. b) Genotoxic effects of triclosan but not 2,4-Dichlorophenol at higher concentrations. 117

Figure 51: No cytotoxicity (a) or genotoxicity (b) of 0.75 mg/L Irgarol 1051 after ozonation. 118

Figure 52: a) No cytotoxic effects of 490 µg/L Terbutryn in HPLC water before and after ozonation. b) Genotoxic effects of Terbutryn before and after ozonation (0 – 195 µg/L O_3) but no genotoxicity using 800 µg/L O_3. 119

Figure 53: UV/H_2O_2 oxidation of 0.1 mg/L AHTN in HPLC water. a) Weak cytotoxic effects after 20 min UV treatment. B) No genotoxic effects before and after UV treatment. ▼ cytotoxic effects 121

Figure 54: Ozonation (10 mg/L O_3) of 0.1 mg/L AHTN in HPLC-water. No cytotoxic (a) and genotoxic effects (b). 121

Figure 55: No cytotoxicity (a) and genotoxicity (b) of HHCB and its oxidation by-products HHCB Lactone at concentrations between 0.05 and 50 µg/L. 122

Figure 56: Cytotoxicity (a) and genotoxicity (b) of 0.1 mg/L HHCB in HPLC-water before (0 min) and after 30 min or 60 min of UV/H_2O_2 treatment (15 W). ▼ cytotoxic effects; n.t. = not tested 123

Figure 57: a) Relative light units (RLU) of the 17ß-Estradiol standard series. b) Relative light units of 0.5 – 50 µg/L HHCB. 124

Figure 58: a) Cytotoxicity and b) genotoxicity of 0.1 mg/L TCPP before (0 min) and after (60 min) ozonation or UV/H_2O_2 treatment. ▼ cytotoxic effects; n.t. = not tested 125

Figure 59: a) No cytotoxicity of untreated and ozonated TCEP in HPLC-water (0.1 mg/L), but cytotoxic effects after 60 min UV/H_2O_2 treatment. b) No genotoxic effects of non cytotoxic TCEP samples. ▼ cytotoxic effects; n.t.=not tested 126

Figure 60: a) No cytotoxicity of 0.1 mg/L TPP in HPLC-water before and after ozonation or UV/H_2O_2 treatment. b) Genotoxicity only before oxidative treatment (0 min) 127

Figure 61: *In vitro* test methods for the determination of endocrine effects adapted from Kase et al. 2009 [71] 131

Figure 62: Schematic overview of the EDA (effect directed analysis) of mixtures published by Werner Brack [234]. 136

Figure 63: Adverse Outcome Pathway proposed by the US EPA for risk assessment based on *in vitro* and *in vivo* methods [236]. 136

II List of Tables

Table 1: Requirements of the waste water treatment plant effluent at the discharge point at 12 °C according to the German "Abwasserverordnung" (concentrations in mg/L) [40]. 25

Table 2: Definition of the Gesundheitliche Orientierungswerte (health assessment values) adapted from the German Umweltbundesamt [73] 37

Table 3: Summary of toxicological test systems required according to corresponding guidelines 41

Table 4: Used test systems 42

Table 5: Cytotoxicity tests required according to DIN EN ISO 10993-5:2009-10 44

Table 6: Selection of *in vitro* tests for the detection of genotoxic effects 45

Table 7: Selection of *in vitro* tests for the detection of mutagenic effects 46

Table 8: Selection of *in vitro* tests for the determination of estrogenic effects 47

Table 9: Cell culture solutions for CHO cells 50

Table 10: Cell culture solution for T47D cells 50

Table 11: Solutions for the PAN I Multitox test 51

Table 12: Solutions for the MTT test 51

Table 13: Solutions for the Alkaline Comet Assay 52

Table 14: Solutions for the Ames Test 53

Table 15: Solutions for the ER Calux 53

Table 16: Steps of calculating the viability of the LDHe test 63

Table 17: Degree of cytotoxicity 65

Table 18: Significance of DNA damage according to the Mann Whitney Test 69

Table 19: Scheme of pipetting for exposure 71

Table 20: Ames MPF® 98/100 quality standards 72

Table 21: ER Calux quality standards 75

Table 22: Chemicals and investigated substances used for oxidation experiments. 76

Table 23: Systems used for oxidation experiments of the tested samples, and their final concentration during toxicological testing. IUTA = Institut für Energie- und Umwelttechnik e.V.; UDE = Institute of Environmental Chemistry, University Duisburg-Essen 77

Table 24: Mass changes of the analytes during HPLC-MS/MS 83

Table 25: Measurement settings for pharmaceutical quantification using the API 3000 system 84

Table 26: Mass transfer of the analytes during GC-MS measurements 86

Table 27: Toxic effects of different amounts of water in exposure medium 89

Table 28: Estrogenicity of 1.5 mg/L Ethinylestradiol dissolved in HPLC-water before and after UV treatment and ozonation 102

Table 29: EEQ values [pM/100 µL] of 0.1 mg/l Bisphenol A in HPLC-water 104

Table 30: Ames test results of 1.4 mg/L Ciprofloxacin in HPLC-water before and after UV/H_2O_2 treatment showing the average number of positive wells. 113

Table 31: Possible mixture effects in regard to biological effects 134

Table 32: Summary of EC_{50} or LC_{50} values of Irgarol 1051 and M1 found in the literature 153

III List of Abbreviations

*	significant
**	very significant
***	highly significant
°C	degree Celsius
2-AA	2-aminoanthracene
2-NF	2-nitrofluorene
3R	Replacement, Refinement, Reduction
4-NQO	4-Nitroquinoline 1-oxide
A. salina	*Artemia salina*
ACN	Acetonitrile
AHTN	1-(3,5,5,6,8,8-hexamethyl-6,7-dihydronaphthalen-2-yl)ethanone
AIF	Allianz Industrie Forschung
AOP	Advanced Oxidation Process
AOX	Adsorbable organic halogen compounds
ATP	Adenosine triphosphate
A-YES	Yeast estrogen screen using the yeast *Arxula adeninivorans*
B	Blank control
BDS	BioDetection Systems
BHK-21	Syrian hamster kidney fibroblasts
BKH	BKH Consulting Engineers
BOD	Biological Oxygen Demand
BOD_5	BOD of five days
c	concentration
$C_4H_2Mg_5O_{14}$	Magnesium hydroxide carbonate
C. dubia	*Ceriodaphnia dubia*
CH	Switzerland
CHO-9	Chinese Hamster Ovary
cm	centimeter(s)
CoRAP	Community Rolling Action Plan
CO_2	Carbon dioxide
COD	Chemical Oxygen Demand
D	Germany
d	day(s)
D. magna	*Daphnia magna*
D. duplex	*Daphnia duplex*
Da	Dalton
DIN	Deutsches Institut für Normung (German Institute for Standardization)
DDD	Defined Daily Dose
DMEM F12	Dulbecco's Modified Eagles Medium
DMSO	Dimethyl sulfoxide
DNA	Deoxyribonucleic acid
DOC	Dissolved Organic Carbon
D. rerio	*Danio rerio*
DTT	Dithiothreitol
e^-	electron

List of Abbreviations

E2	17ß-Estradiol
Eawag	Eidgenössische Anstalt für Wasserversorgung, Abwasserreinigung und Gewässerschutz (Swiss Federal Institute of Aquatic Science and Technology)
EC	European Commission
EC_{50}	half maximal effective concentration
ECHA	European Chemicals Agency
EDA	Effect Directed Analysis
EDC	Endocrine Disrupting Compound
EDTA	Ethylenediaminetetraacetic acid
EDTA	Endocrine Disruptor Testing ans Assessment
EE2	Ethinylestradiol
EEQ	Estradiol Equivalents
EGTA	ethylene glycol tetraacetic acid
ELRA	Enzyme linked receptor assay
EN	European Norm
ENU	N-ethyl-N-nitrosourea
EPA	United States Environmental Protection Agency
EQS	Environmental Quality Standard
ERE	Estrogen Responsive Element
ER Calux	Estrogen Receptor - Chemical Activated Luciferase gene expression
ESI	Electron Spray Ionization
EU	European Union
eV	electron-volt
e.V.	eingetragener Verein
FCS	Fetal Calf Serum
FDA	Food & Drug Administration
Fe(II)	Iron(II)-oxide
Fe^{2+}	divalent iron
Fe^{3+}	trivalent iron
g	gram
G-6-P	Glucose 6-phosphate -
GAC	Granular Activated Carbon
GC	Growth control
GC-MS	Gas chromatography-mass spectroscopy
GOW	Gesundheitlicher Orientierungswert (health assessment value)
H^+	Ptoton
h^+	(electron) hole
h	hour(s)
H. azteka	Hyalella azteca
H_2O_2	Hydrogen peroxide
HAM's	HAM's F12 Nutrient Mixture
HCl	Hydrogen chloride
HClO	Hypochlorite
HCOOH	Formic acid
H_2O	Water
HeLa	human cervical cancer cells

List of Abrreviations

HeLa-9903	HeLA cells transfected with an estrogen receptor and a luciferase gene
HepG2	human hepatocellular carcinoma cells
Hg-Lp UV lamp	Mercury-low pressure UV lamp
HHCB	4,6,6,7,8,8-hexamethyl-1,3,4,7-tetrahydrocyclo-penta[g]isochromene
$HO_2°^-$	Hydroperoxy radical
$HO_3°^-$	Hydrogenozonide radical
HPLC	High performance liquid chromatography
HPLC-MS/MS	High performance liquid chromatography coupled to tandem mass spectrometry
HPRT	Hypoxanthine-guanine-phosphoribosyltransferase
IC_{50}	half maximal inhibitory concentration
IDA	Information Dependent Acquisition
IGF	Industrielle Gemeinschaftsforschung
ISO	International Organization for Standardization
IUTA	Institut für Energie- und Umwelttechnik e.V.
KB	Human cervix carcinoma cells
kg	kilogram(s)
kDA	kilo dalton
kU	Kilo units
kW	Kilo Watt
L929	Mouse fibroblast cells
L	Liter
L. gibba	*Lemna gibba*
L. minor	*Lemna minor*
L. sativa	*Lactica sativa*
LC_{50}	median lethal concentration
LC-MS/MS	Liquid chromatography-Mass spectrometry/Mass spectrometry
LDH	Lactate Dehydrogenase
LDHe	Extracellular Lactate Dehydrogenase
LLE	Liquid-liquid extraction
L.M.P.	Low Melting Point
LOD	Limit of Detection
LOEC	Lowest observed effect concentration
LOQ	Limit of Quantitation
LP-Hg	Low pressure mercury lamp
LUC	Luciferase reporter gene
M	Mole
M1	2-Methylthio-4-*tert*-butylamino-6-amino-*s*-triazine
m^3	cubic meter(s)
m/z	mass-to-charge ratio
mA	milliampere
mbar	millibar
MCF-7	Michigan Cancer Foundation-7; human breast cancer cells
MF	Microfiltration
mg	milligram(s)

List of Abbreviations

$MgSO_4$	Magnesium sulfate
min	minute(s)
mL	milliliter(s)
mm	millimetres
mM	millimole(s)
MMAIII	Monomethylarsonous acid
mmu	Milli Mass Unit
MPF	Microplate formate
mol	mole(s)
MRM	Multiple Reaction Monitoring
MS	Mass spectrometer
MTBE	Methyl tert-butyl ether
MTS	3-(4,5-dimethylthiazol-2-yl)-5-(3-carboxymethoxyphenyl)-2-(4-sulfophenyl)-2H-tetrazolium
MTT	3-(4,5-dimethylthiazol-2-yl)-2,5-diphenyl tetrazolium bromide
MVLN	human breast cancer cell line (MCF-7) stably transfected with the luciferase gene
NaCl	Sodium chloride
NAD^+	Nicotinamide adenine dinucleotide
NADH	Nicotinamide adenine dinucleotide
NADP	Nicotinamide adenine dinucleotide phosphate
NaOH	Sodium hydroxide
NCI-H295R	Human Adrenal Gland carcinoma cells
NEAA	Non-essential amino acids
neg	negative
NF	Nanofiltration
NL	Netherlands
N_{tot}	Total Nitrogen
ng	nanogram(s)
NH_4-N	Ammonium nitrogen
nm	nanometer(s)
NR	Neutral Red
n.s.	not significant
n.t.	not tested
NY	New York
O_2	Oxygen
O_3	Ozone
OD	Optical Density
OECD	Organisation for Economic Co-operation and Development
OH°	Hydroxyl radicals
OH^-	Hydroxide
ONP	Oxidationsnebenprodukt(e): Oxidation by-product(s)
OSPAR	Convention for the Protection of the Marine Environment of the North-East Atlantic
OTM	Olive Tail Moment
ox	oxidized/oxidation
p	p-Value

List of Abbreviations

pM	picomole(s)
P_{tot}	Total Phosphate
PAC	Powdered Activated Carbon
PAN I	Cytotoxicity Test kit
PBS	Phosphate Buffered Saline
PC12	rat pheochromocytoma
pEREtata-Luc	Gene construct containing an estrogen-responsive element linked to a luciferase reporter gene
pH	negative decimal logarithm of the hydrogen ion activity in a solution
Pharm.	Pharmaceutical(s)
PNEC	Predicted no effect concentration
pM	picomole
pos	positive
psi	Pounds per square inch
PTV	Programmed Temperature Vaporizer
QSAR	Quantitative structure-activity relationship
R^2	Coefficient of determination
REACH	Registration, Evaluation, Authorization and Restriction of Chemicals
RLU	Relative Light Units
RO	Reverse Osmosis
ROS	Reactive oxygen species
rpm	rounds per minute
RTG2	Rainbow trut goad tissue cells
rt-YES	rainbow trout Yeast estrogen screen
s	second(s)
S. capricornotum	Selenastrum capricornotum
S9	Liver enzyme mix
SC	Solvent control (negative control)
SD	Standard deviation
SE	Sweden
SEM	Standard Error of Mean
SIM	Selected Ion Monitoring
SPE	Solid Phase Extraction
SRB	Sulforhodamine B
SVHC	Substances of very high concern
t	time
T. platyurus	Thamnocepharus platyurus
T47D	Human breast carcinoma cells
TA98	Salmonella typhimurium strain TA98
TA100	Salmonella typhimurium strain TA100
TAE	Tris-acetate-EDTA
TEER	Transepithelial electrical resistance
TCEP	Tris(2-carboxyethyl)phosphine
TCPP	Tris(2-chloroisopropyl)phosphate
TG	Test Guideline

List of Abbreviations

TiO_2	Titanium dioxide
TK	Thymidine kinase
TK6	human lymphoblasts
TOC	Total Organic Carbon
TOF-MS	Time of flight mass spectroscopy
TPP	Triphenylphosphate
TL	Total LDH control (positive control)
U2-OS	human osteosarcoma cells
UDE	University Duisburg-Essen
UDS	Unscheduled DNA Synthesis
UF	Ultrafiltration
US	United States of America
umu Test	Genotoxicity test using the umuC (<u>u</u>v-<u>mu</u>tagenesis) gene
UV	Ultra violet
UV-C	Ultra violet C
V	Volt
V. fischeri	*Vibrio fischeri*
Vero	African green monkey kidney fibroblasts
VIS	Visible light
W	Watt
WFD	Water Framework Directive
WTK-1	Human lymphoblastoid cells
WWTP	Waste water treatment plant
XPRT	Xanthin-Guanine-Phosphoribosyltransferase
XTT	2.3-bis(2-methoxy-4-nitro-5-sulfopheny)-2H-tetrazolium-5-carboxyanilide inner salt
YES	Yeast Estrogen Screen
µg	microgram(s)
µL	microlitre(s)
µm	micrometer(s)
µM	micromole(s)

1. Summary

The contamination of surface waters with organic micropollutants is a well known problem and increased over the last decades. One reason for their release and consequently their detection in surface waters is their incomplete degradation and therefore insufficient removal during conventional waste water treatment processes. To overcome this problem advanced oxidation processes have been proposed as an additional treatment step, since the formation of highly reactive hydroxyl radicals helps to degrade those substances.

This study was therefore designed to investigate the toxicological properties (cytotoxic, genotoxic, mutagenic and estrogenic) of several substances before and after ozonation or UV/H_2O_2 treatment in either waste water treatment plant effluents or HPLC-water using toxicological *in vitro* methods. These methods are able to identify biological effects of the whole water sample thus the entire complexity of chemicals. In addition cellular based methods will give an overview on possible induction mechanisms which will lead to a manifestation of effects on the organ or even organism level.

In the case of Terbutryn its genotoxicity before treatment was first increased at low ozone dosages (36, 100, and 195 µg/L) but was removed at the highest ozone dosage (800 µg). Similar results have been shown for TPP. Genotoxic effects were only seen before ozonation, but after 60 minutes of ozonation these effects were removed.

The ozonation of Triclosan resulted in the formation of the by-product 2,4-Dichlorophenol. Both substances did not exhibit cytotoxic effects up to a maximum tested concentration of 100 µg/L. Genotoxic effects were only seen for Triclosan starting at a concentration of 10 µg/L. Thus a less genotoxic by-product has been formed during ozonation.

The estrogenicity tests showed that HPLC-water containing Bisphenol A did no longer have estrogenic effects after UV/H_2O_2 treatment whereas neither the ozonation nor the UV/H_2O_2 treatment of Ethinylestradiol for 60 minutes resulted in a complete loss of estrogenic activity. The musk fragrance HHCB in contrast did not show estrogenic effects, neither before nor after ozonation.

For the two beta blockers (Atenolol and Metoprolol), the antibiotics Sulfamethoxazole and Ofloxacin as well as the musk fragrance AHTN neither before nor after ozonation or UV/H_2O_2 oxidation cytotoxic effects were detected. The ozonation of Irgaol 1051 did also not lead to cyto- or genotoxic effects. The results of the oxidation experiments with Bisphenol A, Ciprofloxacin and HHCB demonstrated that formed oxidation by-products differ in regard to toxicity depending on treatment method and water matrix. Cytotoxic effects for all three substances were only detected in HPLC-water after 60 minutes of UV/H_2O_2 but not in waste water treatment plant effluent. In addition no toxicity was detected after the same time of oxidation using ozone. The tested concentrations of TCPP as well as TCEP were not toxic before oxidative treatment. However cytotoxic effects were detected after UV/H_2O_2 treatment whereas the ozonation did not induce cytotoxicity.

In general it can be concluded that ozonation as well as UV/H_2O_2 oxidation are two useful methods for the removal of the here tested substances from the effluents of waste water treatment plants in regard to their degradation and to prevent the formation of toxic oxidation by-products considering the used test methods (MTT Test, PAN I, Alkaline Comet Assay, Ames Test, and ER Calux). However a generalized statement about micropollutants their removal and toxicity is not possible, and needs to be established on a case by case basis since operating conditions as well as the matrix composition vary over time. The successful use of toxicological *in vitro* methods for the detection of biological effects of waste water treatment plant effluents in combination with advanced oxidation processes is confirmed by other studies, e.g. a study performed by the Eawag at the waste water treatment plant in Regensdorf or the PILLS project.

2. Introduction

Water is essential for life and therefore one of the most precious goods on earth. Since the amount of fresh water is limited its quality is the most important factor. The distribution of safe water, the removal of contaminants as well as the protection of the ecosystem and the human health have been a concern for many generations and even more to come.

The occurrence of pharmaceuticals, personal care products and various other anthropogenic substances, so called micropollutants, in fresh water and waste water treatment plant (WWTP) influents is a well known problem [1-5]. Waste waters contain a high diversity of components varying over time and place. These components are microorganisms, metals, inorganic substances, organic substances, and biodegradable organic substances as well as nutrients [6]. For many of these components e.g. nutrients the removal efficiency is high but others like organic substances are released into the environment posing a threat to the environment.

Most of these organic substances are biologically active and thus might have effects on the ecosystems and in the end on the human health. Due to an improvement in analytical methods the detection limits of micropollutants are becoming lower. It has been shown that these substances are not eliminated during waste water treatment and reach surface waters resulting in ng/L to even µg/L concentrations [3, 4, 7-14]. Because of this discovery waste water treatment plants have to face new problems, including new treatment methods as well as new methods to control the quality of WWTP effluents. Hereby the contaminated WWTP effluent might pose a threat to human health and lead to problems in drinking water treatment since e.g. pharmaceuticals are designed to work at low concentrations. Therefore appropriate methods need to be developed. These methods need to take into account the complete degradation of water contaminants to harmless molecules. Due to the variety of those contaminants there is no common treatment until now. The use of advanced oxidation processes (AOP) has been widely studied and AOP have been shown to be a promising method although in some cases a downstream filtration (e.g. through activated carbon or sand) is needed [8, 11, 14-23]. It has also been shown that an incomplete degradation promotes the formation of oxidation by-

Introduction

products that differ in structure and function from their parent compound and might therefore also have different toxicological effects on the environment [24, 25]. Therefore methods which are able to detect biological effects of organic micropollutants are required. Toxicological *in vitro* methods are helpful in the detection of these effects since there is a great variety of tests available with different endpoints which allow the identification of a complex mixture of substances in waste water treatment plant effeluents as a precursor for the manifestation of effects on the level of organs or organisms [26, 27].

2.1. Micropollutants in waste water treatment plant effluents and surface water

2.1.1. Guidelines for water quality control

Although a variety of regulations for the safety and cleanliness of water systems have already been established in regard to the water quality on a national basis, the implementation of the Water framework directive (WFD) by the European Union is a step toward a more sincere water quality management internationally. This directive requires the monitoring of biological, chemical, and quantitative parameters for an environmental risk assessment. As part of the WFD 33 substances have been chosen as priority substances [28], and environmental quality standards have been regulated in an EU directive (2008/105/EG) to prevent short term as well as long term effects of priority substances on the environment with maximum concentrations between 0.004 and 50 µg/L depending on the substance [29]. In addition to the European regulations other substances which are not regulated in these directives can be regulated by each member state. The German "Oberflächengewässerverordnung" (Regulation for the protection of surface waters) regulates a variety of parameters in accordance with the two above mentioned directives. Priority as well as other substances need to be measured 4 – 13 times each year depending on their classification [30].

2.1.2. Emission of micropollutants

The routes by which micropollutants reach the water system are adverse (Figure 1). Pharmaceuticals are mainly introduced via the fecal-oral-cycle or by the disposal through the plumbing system finally reaching the waste water treatment plant. But not only the drug itself reaches the water systems, its metabolites are also introduced. After the ingestion of drugs only a part of them gets completely metabolized resulting in an input of residues as well as the original substance into the water system via excretion. This is true for both, human and veterinary pharmaceuticals. The same route can also be applied for personal care products which are used in the everyday life. Shampoos, soaps, perfumes etc. containing e.g. musk fragrances, are used on a daily basis and are also flushed down the drain reaching the water system. Biocides like herbicides, algicides, and insecticides are mainly introduced into the water by runoff from areas where they have been applied. Other sources of introduction are industrial and hospital waste waters which usually contain a high amount of organic substances. In addition to these common routes accidental spills or the intentional illegal disposal of chemicals etc. also contribute to the presence of anthropogenic substances in the water system. Thus it is either a direct input into the surface water or an input through waste water treatment plant effluents after an incomplete removal.

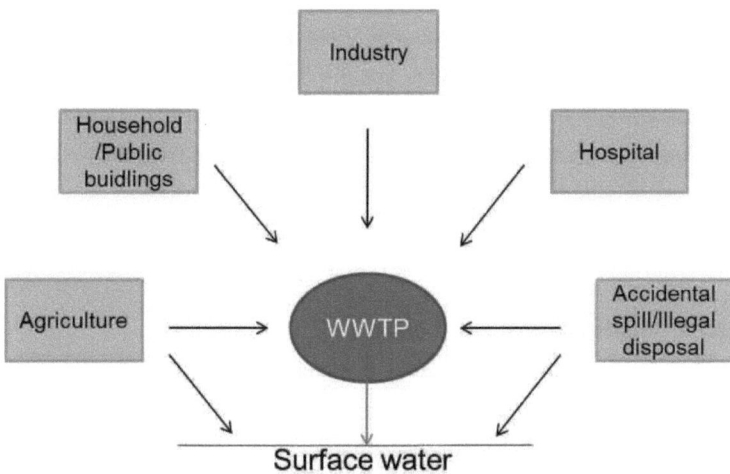

Figure 1: Routes of micropollutant contamination of surface waters

Introduction

2.1.3. Substances detected in waste water treatment plant effluents

The group of organic micropollutants detected in waste water treatment plant effluents is composed of a variety of substances including pharmaceuticals, pesticides, personal care products, and other chemicals. Thus substances with different fields of application and a huge variety of properties are detected in WWTP effluents and receiving surface waters [31, 32]. The presence of pharmaceuticals is due to their ubiquitous use with 8280 drugs allowed in human medicine in Germany [33], as well as their high persistence. In 2010 the volume of defined daily doses (DDD) of the 30 most prescribed pharmaceuticals increased of 3.5 % to an amount of 35384 million DDD which reflects the trend of the last years [34] although the amount of DDD for some pharmaceuticals is decreasing. However this number only gives the amount of pharmaceuticals with the need of a prescription (6288 drugs) not taking into account over-the-counter drugs (1992 drugs), which have to be added to this number. Most of the pharmaceuticals commonly detected are also part of the WHO Model List of Essential Medicines, thus a bigger decline in its uses is improbable [35]. A literature and database review by Bergmann et al. (2011) demonstrates the large amount of pharmaceuticals present in water systems as well as the lack of information on their behavior in the environment [12]. In addition to human pharmaceuticals veterinary drugs need to be considered since they also reach waste water treatment plants and surface waters. Musk fragrances are another group of substances frequently detected in surface waters. They are constituents of personal care products and are thus also used on a daily basis. Like pharmaceuticals they are not easily biodegradable which leads to their detection in WWTP effluents and subsequently in surface waters and sediments [36-38].

2.1.4. Waste water quality control

Due to the presence of a variety of substances in the influents of waste water treatment plants the treatment efficiency of the plants needs to be surveyed to prevent the contamination of receiving surface waters. The quality control of waste water treatment plant effluents in Germany is usually determined by physical and chemical parameters. Sum parameters like the COD (chemical oxygen demand),

Introduction

BOD (biological oxygen demand), TOC (total organic carbon), DOC (dissolved organic carbon), and the conductivity are determined. Group parameters describing substances with similar chemical properties, e.g. total nitrogen, are also determined. In addition characteristics like the color, turbidity and smell are recorded [39]. All these parameters are regulated in the "Abwasserverordnung" (Waste Water Ordinance) which defines the limit values allowed to release the waste water treatment plant effluent into surface waters (Table 1) taking into account different industrial waste waters. A variety of anions (e.g. fluoride, nitrogen compounds, phosphate, sulfate) and cations (e.g. aluminium, arsenic, copper, mercury) as well as adsorbable organic halogen compounds (AOX) and lipophilic compounds need to be determined. Besides these many other parameters need to be monitored. In regard to biologically active substances two *in vitro* methods, the umu-test for the determination of mutagenic effects and the luminescent bacteria inhibition test for the detection of cytotoxic effects are recommended in the German Abwasserverordnung. However these methods are based on bacterial responses and effects might differ when tested with eukaryotic cells. The *in vivo* methods listed in the Abwasserverordnung include the fish-egg test and the testing for toxicity in daphnia and algae, performed according to the corresponding DIN rules [40].

Table 1: Requirements of the waste water treatment plant effluent at the discharge point at 12 °C according to the German "Abwasserverordnung" (concentrations in mg/L) [40].

	Chemical oxygen demand (COD)	Biological oxygen demand (BOD)	Ammoniac nitrogen (NH_4-N)	Total nitrogen (N_{tot})	Total phosphate (P_{tot})
Class 1	150	40	-	-	-
Class 2	110	25	-	-	-
Class 3	90	20	10	-	-
Class 4	90	20	10	18	2
Class 5	75	15	10	13	1

Class 1: < 60 kg/d BOD_5; Class 2: 60 – 300 kg/d BOD_5; Class 3: 300 – 600 kg/d BOD_5; Class 4: 600 – 6,000 kg/d BOD_5; Class 5: > 6,000 kg/d BOD_5

Introduction

Depending on their classification, waste water treatment plants need to meet certain criteria. The waste water treatment plants are classified according to the BOD_5 value. This value describes the amount of oxygen needed for the aerobic degradation of organic materials by microorganisms. Thus it gives a measure for the degree of pollution. It is an approximate measure for the amount of oxidized organic substances after five days because at that time most of the organic matter is supposed to be degraded by microorganisms [6]. The BOD_5 therefore is a value of the population equivalent describing the pollution load.

2.1.5. Common waste water treatment

The first ideas of waste water treatment reach back to ancient times. However the development of techniques used during waste water treatment nowadays has been started in the late 19th century followed by a steady development of new techniques and ways of treatment [41].

A common waste water treatment plant consists of a pre-treatment and three subsequent treatment steps (Figure 2). Before the primary treatment is applied the waste water is commonly subjected to the pre-treatment (mechanical treatment) where materials like branches, leaves or litter are mechanically removed by a rack. During primary treatment, the incoming pre-treated waste water reaches a tank where it remains for some time to allow the settlement of bigger solid particles (e.g. sand). Dissolved as well as suspended organic matter which was not removed during primary treatment is removed during secondary treatment (aerobic-biological treatment). This step includes a basin where microorganisms degrade the remaining organic matter by the addition of oxygen for an aerobic environment. The best possible outcome of this treatment step is a complete degradation of the organic matter as well as an oxidation of the inorganic matter. The last step before the water is finally released into the environment is the tertiary treatment where the activated sludge originating from secondary treatment is removed by deposition. Part of the activated sludge is transported back into the secondary treatment basin whereas the

Introduction

other part is transferred to a final sludge treatment which can be done by digestion or incineration before it gets deposited. In addition to this a further treatment (e.g. disinfection, sand filtration) can be applied according to the further use of the waste water treatment plant effluent but is not mandatory. An important part of the waste water treatment process is the removal of nutrient salts. The removal of nitrogen compounds (nitrification/denitrification) is linked to the biological treatment of the secondary treatment. Phosphorous compounds can be removed by flocculation with a subsequent filtration [39].

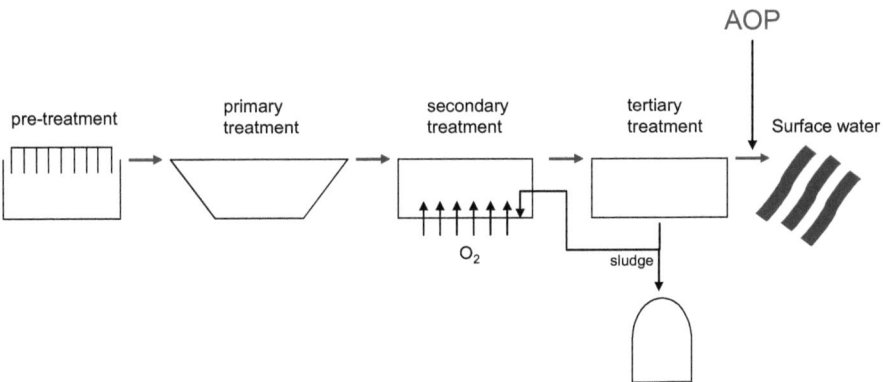

Figure 2: Schematic of a waste water treatment plant. AOP = Advanced Oxidation Processes

However as described before, not all organic substances (e.g. pharmaceuticals, personal care products) are removed during the whole treatment process and a fourth treatment step is therefore needed to ensure a good water quality according to the WFD and national regulations before the effluent is released into receiving waters. A variety of possible methods to be applied as a fourth treatment step are discussed below.

Introduction

2.1.6. Advanced waste water treatment

2.1.6.1. Filtration

The use of membranes for water purification processes has been established in the early 20th century with a steady improvement using ceramic or polymeric materials with defined pore sizes (0.05 – 5 µm). The principle of a membrane is the filtration based on the properties of some substances to pass the membrane without being altered. If a substance will pass the membrane is determined by the size of the membrane pores thus the pore size is selective. According to the pore size four membrane processes can be distinguished (Figure 3). These are reverse osmosis (RO; pore size < NF), nanofiltration (NF; pore size < 2 nm), ultrafiltration (UF; pore size 2 – 50 nm), and microfiltration (MF; pore size > 50 nm) listed according to the pore sizes from small to large [42].

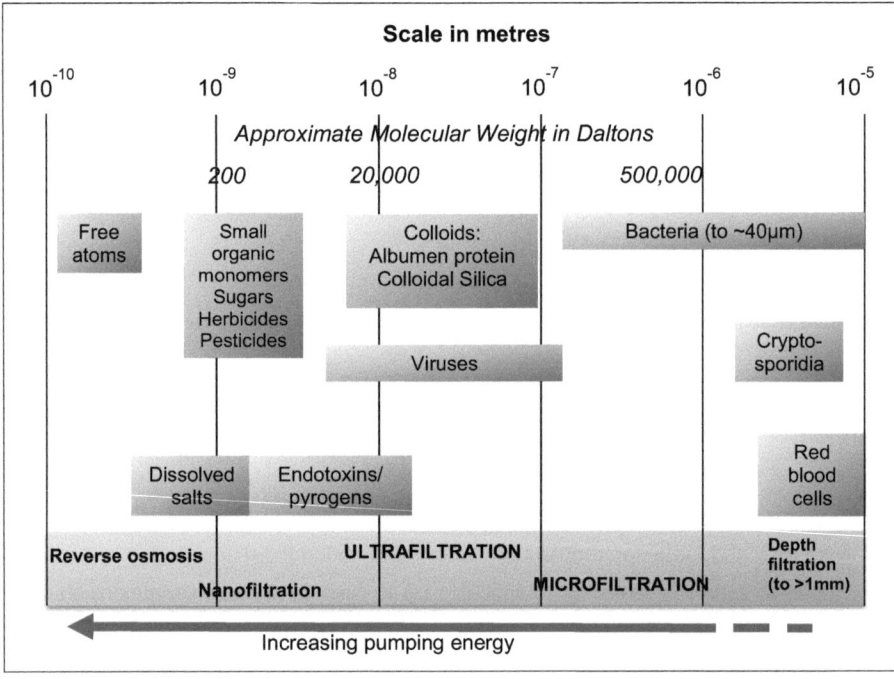

Figure 3: Membrane separation processes overview [42].

Introduction

Reverse osmosis is a technique commonly used for the desalination of seawater by separating the water molecules and the ions through a membrane under pressure (20 – 100 bar) [31]. This method is also thought of as an additional treatment step to remove remaining micropollutants after waste water treatment. However there are some limitations. Organic substances might be dissolved in the membrane remaining there for a certain time before they are released. Thus also the water is thought to be clean, at some point it might be polluted again. Another limitation is the property of organic substances to penetrate the membrane more easily than the water molecules resulting in no removal at all [31, 43]. Nanofiltration is used for the separation of dissolved organic matter (> 200 g/mol), divalent cations and anions. It is therefore often used for the softening of water. During ultrafiltration viruses and colloids are rejected by the membrane whereas the microfiltration technique having the biggest pore size is only able to separate suspended solids, e.g. bacteria and bigger materials from the water [42]. This defined pore size is at the same time the limitation of the filtration process because the pores might easily clog or biofilms are formed which leads to a high energy consumption [39, 43]. Other facts like biofouling also limit the time of usage of the membranes. These facts in addition with high costs and the operating expenses limit the use of especially reverse osmosis but also the other three filtration techniques as a fourth treatment step.

Although oxidative treatment methods have been proven successfully in the complete removal of micropollutants a further treatment using filtration techniques might be useful especially for the removal of oxidation by-products. One of these filtration methods is the sand filtration. Sand filters are commonly used for the purification of water, both during drinking and waste water treatment. While sand filters are mainly used for disinfection purposes of drinking water and the removal of odor and taste they are used as an additional step during waste water treatment as a filter for the adsorption of remaining organic matter. The final report of the Eawag (Swiss Federal Institute of Aquatic Science and Technology) regarding the ozonation of treated waste water showed that the ozonation of the WWTP effluent already resulted in a high removal efficiency, however after sand filtration the removal rate was even higher especially with regard to formed by-products [14, 44].

Introduction

2.1.6.2. Sorption: Activated Carbon

Activated carbon describes a class of carbon materials produced synthetically from materials like wood, nut shells or coal. One advantage of activated carbon is its highly porous surface and therefore big area which allows reactions with substances [43, 45]. The porous surface is produced by drying and heating the carbon material using air, steam or carbon dioxide. A subsequent heating step using oxidation gases or CO_2 at temperatures above 800 °C is applied for activating the surface area by increasing it [46]. Although activated carbon is used in many fields (e.g. product purification) its main application is during environmental processes for a pollution control. Two types of activated carbon are available: Granular Activated Carbon (GAC) and Powdered Activated Carbon (PAC). GAC is composed of crushed granules originating from coal or shell and the particles are between 0.2 and 5 mm. PAC in contrast is made from wood with a particle size of 15 to 25 µm. While PAC can either be added to the activated sludge or in a separate tank, GAC is always used as a separate treatment step.

Besides the use in drinking water treatment activated carbon is also used during waste water treatment. The first application of activated carbon for waste water treatment was in California in 1965 in a municipal plant. Nowadays activated carbon is mainly used for the treatment of industrial waste waters [45].

During waste water treatment activated carbon is used as a material to adsorb dissolved organic matter because of the high adsorbance capacity [47-50]. It has also been proven to remove the toxic activity after ozonation [24]. Due to its high costs e.g. through the processing for reuse, it is usually only applied when no other treatment is successful or as a last treatment step when most of the micropollutants have already been removed and only substances which are not biodegradable are present [45].

Introduction

2.1.6.3. Advanced Oxidation Processes (AOP)

Since organic micropollutants are not removed during the different steps of common waste water treatment and some of them are highly persistent another fourth treatment step is needed. Currently it is being discussed which technique would be the best to eliminate these substances and thus there are a variety of requirements. From an economic point of view the fourth step needs to be easy to apply and most efficient with low costs. Another fact that needs to be considered is the minimization of an environmental risk through the use of an additional treatment step. In this context advanced oxidation processes (AOP's) are believed to be the best possible method and thus different methods have been developed and successfully applied [51-55]. However, the oxidation of micropollutants has some disadvantages. It has been shown for drinking water treatment that the presence of e.g. bromide results in the formation bromate, a suspected carcinogen, through disinfection processes [56, 57]. To avoid this disadvantage of oxidation processes, the chosen AOP is of great importance and might vary depending on the composition of the waste water.

Advanced oxidation processes are defined as processes which initiate the formation of hydroxyl radicals (OH°) in such amounts that they are able to support the water purification [58] by degrading water contaminants. Due to their highly reactive (second order rate constants $10^6 - 10^9$ M^{-1} s^{-1}) and mostly unselective properties as well as their high oxidation potential (2.8 V) they react with the organic matter resulting in an oxidation of the pollutants [59, 60]. Another similarity of all possible AOP's is the need of a pre-treated wastewater with a low chemical oxygen demand (COD; ≤ 5 g/L) since the AOP alone is not sufficient in removing the organic matter (Figure 4). A high organic load would increase the amount of oxidant needed and would not only raise the treatment costs but would also lower the efficiency of treatment. The best possible outcome after an oxidative treatment is the complete mineralization of the micropollutants to CO_2, H_2O or an inorganic product. An incomplete mineralization only promotes the formation of by-products [51]. The degradation process itself is mainly unknown for many substances and is specific for each applied method.

Introduction

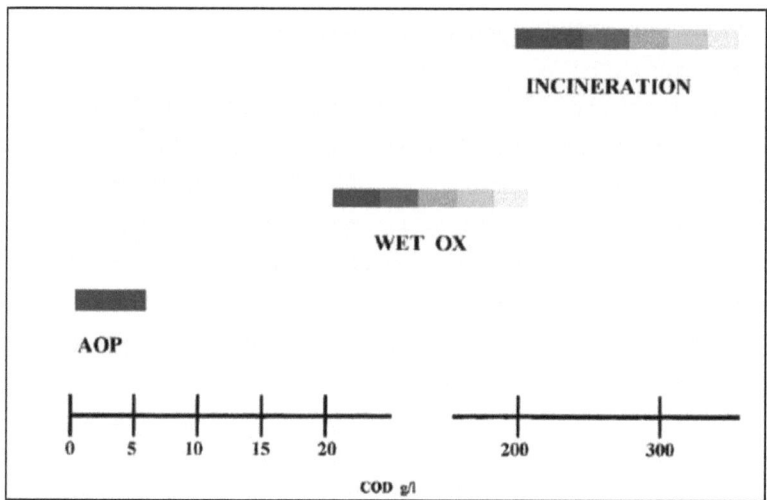

Figure 4: Suitability of water treatment technologies according to the chemical oxygen demand (COD in g/L) [51].

2.1.6.3.1. Ozone

Apart from chlorine, ozone is nowadays one of the most used disinfectants during water treatment processes. Besides its beneficial effects during drinking water treatment, e.g. removing taste, color, and odor, as well as the elimination of microorganisms and humic substances, the use of ozone still has some disadvantages. Although ozone is known to be unstable in water it easily undergoes reactions with water compounds resulting in the formation of by-products. The mechanisms already described for drinking water treatment can also be applied for the use of ozone as an additional step in waste water treatment. In contrast to drinking water treatment the focus of the ozone usage during waste water treatment is the removal of organic micropollutants and not on disinfection. In addition the resulting by-products might differ from those known for drinking water depending on the much more complex water matrix.

The ozonation itself can take place in two ways with different reactions. The reactions can either occur through a direct reaction by the ozone molecule itself or an indirect reaction through the formation of OH° (Figure 5), thus leading to different oxidation by-products and different reaction rate constants (direct = $1.0 - 10^6$ M^{-1} s^{-1},

Introduction

indirect = $10^8 - 10^{10}$ M^{-1} s^{-1}). Another difference of the two reaction pathways is the way of the oxidative attack. A direct ozonation by O_3 is selective and compounds with unsaturated bonds (e.g. aromatic compounds) are mainly attacked. The higher the electron density the faster the reaction takes place. Indirect reactions are initiated through the presence of OH^- ions. At the end of the reactions hydroxyl radicals are formed which then subsequently react with the micropollutants thus oxidizing them.

$$O_3 + OH^- \rightarrow O_2^{-\circ} + HO_2^\circ \quad \longrightarrow \quad O_3 + O_2^{-\circ} \rightarrow O_3^{-\circ} + O_2$$

$$\downarrow$$

$$HO_3^\circ \rightarrow O_2 + (OH^\circ) \quad \longrightarrow \quad O_3^{-\circ} + H^+ \leftrightarrow HO_3^\circ$$

Figure 5: Reaction pathways of the OH° (encircled) formation by ozone [48]

Both reactions can take place simultaneously but depending on the properties of the water (pH, temperature) one of the reactions is favored. A high pH for example results in an oxidation through OH° whereas a low pH favors the direct ozonation. This also indicates that depending on the water matrix (DOC content) one of the reactions dominates. In addition ozone is not specific for organic miropollutants thus inorganic compounds e.g. iron or nitrogen compounds might also be oxidized [61] therefore a complete elimination of these inorganic compounds before the use of ozone is advantageous.

2.1.6.3.2. UV treatment with and without H_2O_2

UV light is a radiation with a wavelength of 4 – 400 nm and invisible to the human eye. In order to be oxidized by UV light, substances present in the water need to be able to absorb this radiation. They mainly absorb UV light at wavelengths between 200 and 280 nm [60] thus the range of the UV-C radiation which is 100 – 280 nm [62]. During the treatment of wastewater, low pressure mercury lamps (LP - Hg) are mainly used with a wavelength of 253.7 nm. However the sole use of UV light in regard to the oxidative potential is limited since the substances have to absorb the

Introduction

UV light to become oxidized. The addition of H_2O_2 resolves this limitation by the formation of highly active OH radicals (OH°) which then react with the micropollutants in an unselective way.

Therefore UV treatment processes can either include the direct photolysis with the sole use of UV radiation (Figure 6a) or the use of a combination with OH° driven reactions through the additional use of H_2O_2 resulting in an indirect oxidation reaction (Figure 6b). The OH° formation during UV/H_2O_2 treatment is based on the photolysis of H_2O_2 [60]. Due to the two varying reaction pathways different oxidation by-products might be formed.

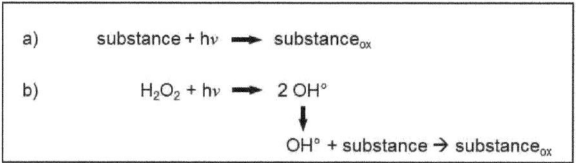

Figure 6: Oxidation and reaction pathways of the UV treatment without (a) and with (b) H_2O_2

However this method is also limited by the turbidity of the WWTP effluent due to effects of the water matrix. This can result in an incomplete degradation because big molecules are only partially oxidized and the treatment duration would therefore increase in order to result in a complete degradation of the micropollutants to CO_2 and H_2O.

2.1.6.3.3. Fenton and Fenton like reactions

The Fenton reaction was first described in the 1890's by Henry John Horstman Fenton. Fenton detected the oxidation of tartaric acid in the presence of H_2O_2 and iron ions. The use of a mixture composed of H_2O_2 and ferrous iron thus enables the oxidation of organic molecules and the use during waste water treatment [63].

Introduction

During the Fenton reaction, iron-compounds acting as the catalyst are reduced resulting in a free iron ion (Fe(II)) which will react with H_2O_2 thus leading to the formation of $OH°$ radicals (Figure 7) [21, 64].

$$Fe^{2+} + H_2O_2 \rightarrow Fe^{3+} + OH^- + OH°$$

Figure 7: Fenton reaction

Although this reaction already results in the formation of $OH°$ the additional use of irradiation increases the degradation rate due to the photoreduction of the Fe^{3+} to Fe^{2+} ions. Subsequently the Fenton reaction can take place where the formed Fe^{2+} ions are then able to produce more $OH°$ in the presence of H_2O_2 (Figure 8) [65]. This reaction mechanism is then called Photo-Fenton reaction.

$$Fe^{3+} + H_2O + h\nu \rightarrow Fe^{2+} + H^+ + OH°$$
$$Fe^{2+} + H_2O_2 \rightarrow Fe^{3+} + OH^- + OH°$$

Figure 8: Photo-Fenton reaction

These processes have already been shown to be sufficient in the removal of organic micropollutants from waste water treatment plant effluents [21, 65-69]. However the applicability due to the complex reaction needs as well as the need for a low pH limits its use for the treatment of waste water treatment plant effluents.

2.1.6.3.4. Photocatalysis

Another possible treatment method includes the use of TiO_2 nanoparticles acting as a semiconductor. These particles are irradiated in the presence of oxygen finally leading to the formation of $OH°$ (Figure 9) including redox reactions of adsorbed water, oxygen and other molecules. The use of the UV radiation results in the

Introduction

formation of an electron-hole pair subsequently followed by a transfer of the electron leading to the formation of OH° radicals [21, 64, 70].

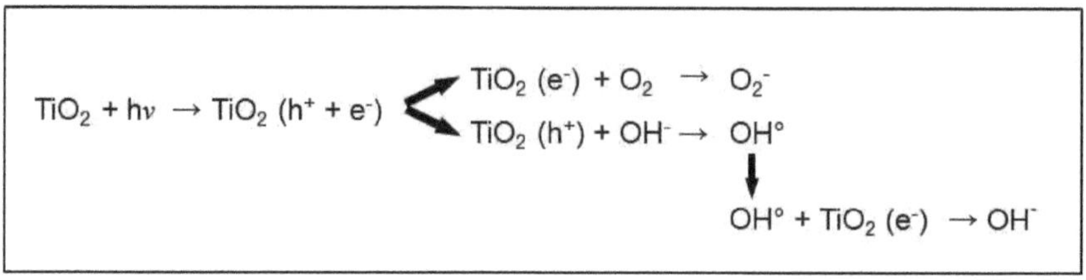

$$TiO_2 + h\nu \rightarrow TiO_2 (h^+ + e^-)$$
$$TiO_2 (e^-) + O_2 \rightarrow O_2^-$$
$$TiO_2 (h^+) + OH^- \rightarrow OH°$$
$$OH° + TiO_2 (e^-) \rightarrow OH^-$$

Figure 9: Reaction pathway of the TiO$_2$ photocatalysis

However, despite all these possible methods only few of them are feasible. Photocatalysis, as well as membrane and Fenton processes are limited due to their quite complicated requirements as well as the high costs when applying these methods. In contrast processes like ozonation and UV/H$_2$O$_2$ treatment are more efficient in regard to costs, applicability and the removal of micropollutants through OH° formation. All methods also might result in the formation of oxidation by-products due to an incomplete degradation which should be prevented by finding the most applicable treatment setting.

2.2. Toxicological testing for the determination of biological effects of micropollutants

The use of toxicological methods helps to identify substances as well as the conditions (e.g. concentration, time of exposure) which lead to biological effects in a dose dependent manner. In fact not one single test is sufficient in predicting the possible risk, thus many tests need to be performed in order to perform a risk assessment. For example Kase et al. (2009) proposed a multistep process for the investigation of endocrine or reproductive effects suggesting the combination of different *in vivo* and *in vitro* methods which allow the detection of toxicological effects at the molecular, cellular as well as organism level. These tests should be able to identify effects selectively and mode of action based and an extrapolation of these

Introduction

effects to organisms or population effects should be possible [71]. Looking at the process of placing a pharmaceutical on the market it can be seen that it is a multistep process including a variety of chemical, physical and toxicological test. However this multistep process is only applicable for the testing of chemicals etc. but not for water samples. As can be seen in the German "Abwasserverordnung" the quality of waste water is mainly determined by chemical parameters [40]. In fact with a steady improvement in analytical methods the concentrations of substances detected in water systems are becoming lower. However the sole detection, respectively the quantification, of a substance does not give any information on its biological effects. In addition, the detection of a single substance does not give any information about accumulative effects occurring in a water sample. These effects in contrast can be detected and determined using biological test systems. Therefore the establishment of toxicological methods for the use in water testing is important [72]. This becomes even more important in regard to the definition of limit values especially with the focus on the human health and an exposure as low as possible. As a matter of fact, in 2003 the German Umweltbundesamt recommended the use of "Gesundheitliche Orientierungswerte" (GOW, health assessment value) for the evaluation of substances present in drinking water when only little or no information regarding their toxicity is available (Table 2). This GOW sets a limit of the highest concentration of a substance allowed in drinking water and is based on the available data on its genotoxic and neurotoxic effects [73].

Table 2: Definition of the Gesundheitliche Orientierungswerte (health assessment values) adapted from the German Umweltbundesamt [73]

Available information about substance	*GOW [µg/L]*
Predominantly no *in vitro* genotoxicity and carcinogenicity of the substance, but otherwise no significant ecotoxicological data is available	≤ 0.3
No verifiable genotoxicity (see above). In addition significant *in vitro* and *in vivo* data on oral neurotoxicity and on its germ cell damaging effects are available. These data do not lead to values below 0.3 µg/L.	≤ 1
The substance is neither genotoxic, nor germ cell damaging or neurotoxic (see above). In addition significant *in vivo* data of at least one study on the subchronic-oral toxicity is available. These data do not lead to values below 1 µg/L.	≤ 3

Introduction

Pharmaceuticals have been designed to be biologically active in order to induce a beneficial effect on humans. However at the same time they might lead to adverse effects when exposed at higher doses or a mixture of those bioactive substances as present in water systems. It has already been shown that certain contaminants are linked to the development of diseases e.g. cancer as a result of a heavy metal exposure. But besides cancer other health aspects e.g. effects on the nervous or immune system as well as the endocrine or reproductive system are of concern. Because these systems are important for the basic functions of an organism it is important to identify environmental contaminants and their possible threat to the whole environment. This includes the evaluation of a human risk potential and subsequently the definition of limit values (e.g. GOW) based on results gained from toxicological studies. Advances in toxicological testing approaches are therefore needed in order to establish a reliable human risk assessment concerning environmental contaminants. This includes a chemical characterization of those contaminants in combination with toxicological tests based on human or in general eukaryotic cell lines (Figure 10) [74].

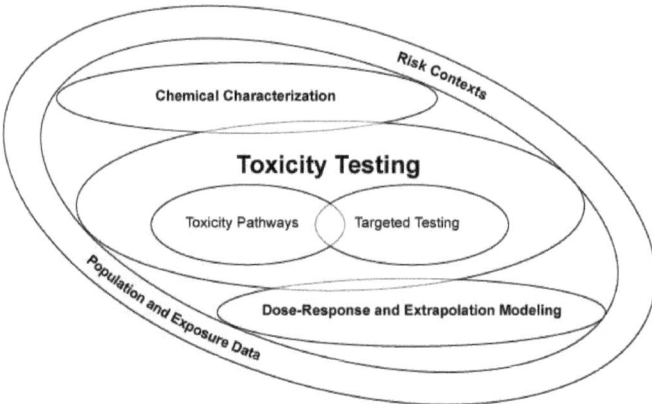

Figure 10: Testing strategy proposed by the Committee on Toxicity Testing and Assessment of Environmental Agents [74]

Introduction

Toxicological *in vitro* methods using cell culture based systems are thereby important first step approaches to identify possible risks at the molecular level to gain information on outcomes at the organisms or population level (Figure 11) [26].

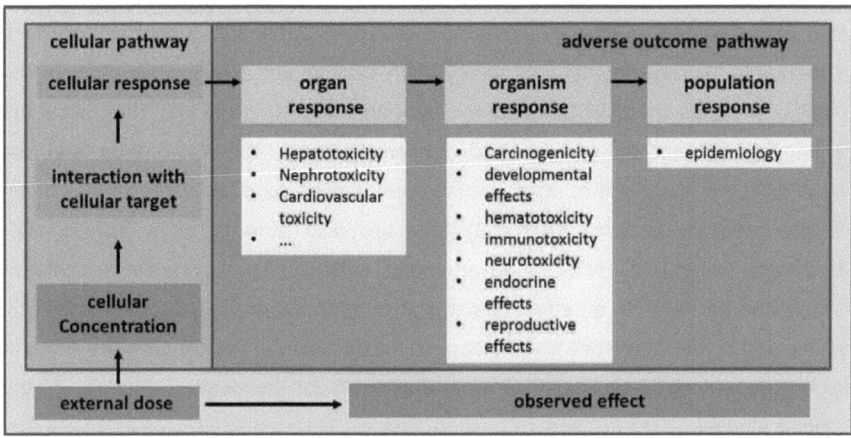

Figure 11: Bioanalytical tools for the assessment of human health risks adapted from Escher et al. 2011 [26]

The field of toxicology in general deals with the detection and identification of substance properties in regard to their potential to cause harm. The use of toxicological *in vitro* methods for the testing of chemicals is widespread because they can be used to gain data on the mechanisms of cell damage which in fact can be transferred to humans. Especially with the focus on the REACH regulation (**R**egistration, **E**valuation, **A**uthorization and Restriction of **Ch**emicals) by the EU council the minimization of animal testing through the use of *in vitro* methods is supposed [75].

On the basis of various of initiatives to reduce the number of animals in toxicity testing [76-79] in regard to the "3R" principle (Replacement, Reduction, Refinement) *in vitro* methods should be established and performed. A variety of studies have been published proving a good correlation between *in vivo* and *in vitro* data but a combination of both would be most favorable [80, 81]. *In vivo* studies require a large amount of testing animals and range from short term acute studies to long term

Introduction

chronic studies sometimes including several generations. Limitations of these studies are the high efforts and costs. However the advantage is their ability to e.g. detect systemic influences and healing processes. In addition, the kinetics (intake, distribution, metabolization, excretion) of a substance including a metabolization after exposure can only be measured *in vivo*. *In vitro* methods in contrast, are fast, non-complex screening methods with a high throughput, a lesser variation of the results, and thus a higher potential for standardization (Figure 12). They allow the testing of different substances and concentrations at the same time and the identification of the mode of action, thus lowering efforts and costs. With the addition of liver enzyme mixtures (e.g. S9) they are also able to show effects after metabolization [82-84]. Another advantage of cellular based systems is that their results can be used for an effect description for different organisms. Although both, *in vivo* and *in vitro,* methods can be applied for the testing of water samples it has to be decided which tests would be most applicable for the desired outcome of the specific studies.

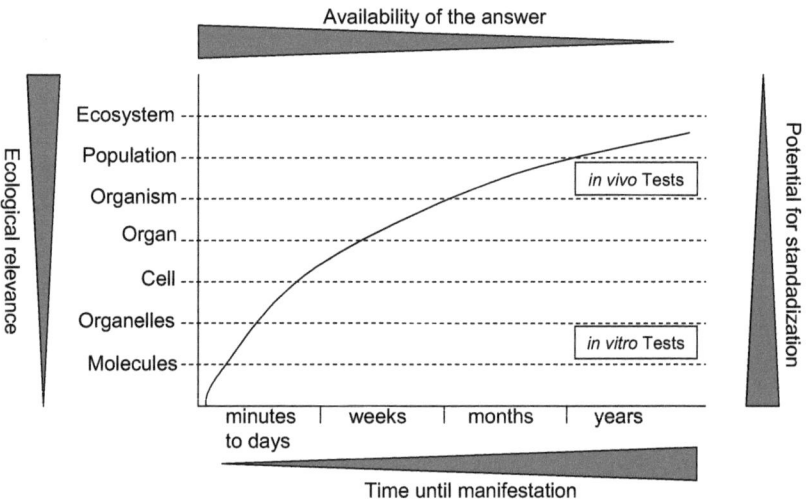

Figure 12: Level of toxicological effect and time until manifestation of effects in regard to the method, adapted from Kase et al. 2009 [71].

Introduction

Since the variety of micropollutants as well as their biologic effects are adverse a battery of different toxicological methods need to be applied to gain an understanding on their mode of action. In combination with chemical techniques an effect-directed analysis is then possible. With respect to existing guidelines and toxicological endpoints effects like general cell damages (cytotoxicity), DNA damages (genotoxicity), heritable DNA damages (mutagenicity), and effects on the endocrine system (e.g. estrogenicity) should be observed when applying *in vitro* methods. The methods used for determining biological effects therefore need to be selective and sensitive to detect effects resulting from low concentrations of substances usually found in water bodies.

Comparing the different regulations which are applied for water treatment in Germany in regard to the methods required to determine the water quality only the Abwasserverordnung (waste water guideline) requires different toxicological tests. In addition only biota studies need to performed according the the german surface water guideline as well as the Water Framework Directive whereas all other guidelines for water quality control solely include chemical analyses (Table 3) [28, 30, 40, 85, 86].

Table 3: Summary of toxicological test systems required according to corresponding guidelines

Test Method	Surface water	Drinking water	Groundwater	Waste water	Water framework directive
Fish egg test	-	-	-	yes	-
Daphnia test	-	-	-	yes	-
Algal growth inhibition test	-	-	-	yes	-
Luminescent bacteria test	-	-	-	yes	-
Umu test	-	-	-	yes	-
Biota studies	yes	-	-	yes	yes

Introduction

A battery of four different biological *in vitro* tests was therefore used in this study to determine the behavior of micropollutants in waste water treatment plant effluents based on their cytotoxic, genotoxic, mutagenic, and estrogenic effects. Displayed in Table 4 are the used test methods according to the mode of action.

Table 4: Used test systems

Cytotoxicity	MTT Test	Mitochondrial activity
	PAN I: • LDHe • XTT Test • Neutral Red • Sulforhodamine B	Membrane integrity Mitochondrial activity Lysosomal activity Total protein content
Genotoxicity	Alkaline Comet Assay	Single strand and double strand DNA breaks, alkali labile sites
Mutagenicity	Ames Test	Frameshift (TA98) and base pair substitution mutations (TA100)
Estrogenicity	ER Calux	Human estrogen receptor activation

These *in vitro* tests were chosen since the focus was laid on the effects of micropollutants and their oxidation by-products on the human health and considering the requirements of the GOW concept which also includes the use of *in vitro* data for the establishment of limit values. The tests were also chosen with regard to their possible application for screening purposes due to their ease of applicability, low costs and high throughput.

This battery of tests was then applied to the non-oxidized as well as to the oxidized samples after ozonation or UV ± H_2O_2 treatment and first tested for cytotoxic effects. In case of no cytotoxic effects the samples were then subsequently tested for genotoxicity. The occurrence of mutagenic or estrogenic effects was further on only investigated when the substances were known or supposed to induce such effects (Figure 13). Chemical analyses (grey box) of the same samples were also performed by collaborating groups but not as a part of this thesis.

Introduction

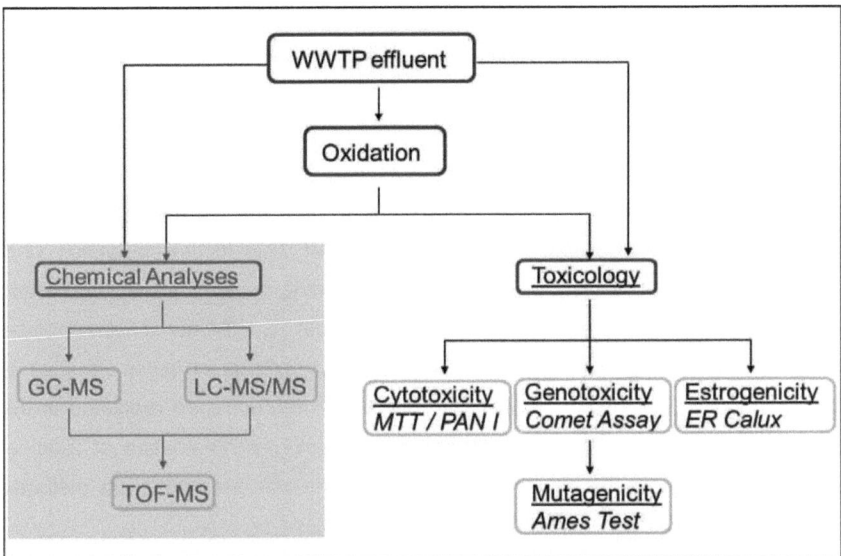

Figure 13: Schematic of toxicological and chemical (grey box) testing as part of this project. Chemical analyses have been performed by the working groups of the IUTA e.V. or the University of Duisburg-Essen.

Introduction

2.2.1. Cytotoxicity

A substance is defined as cytotoxic when cellular functions are disturbed or if it induces cell death. Tests to determine the cytotoxic effects of a single substance or a mixture are used for the investigation of damages to basic cellular functions e.g. membrane integrity or mitochondrial activity. These results can then be further used for the investigation of other toxic effects (e.g. genotoxicity, mutagenicity and estrogenicity) using sub-cytotoxic concentrations [84]. Due to the complexity of a cell, the mode of action of chemicals differs depending on their affinity for certain structures (e.g. membrane, mitochondria, lysosomes). So far no *in vitro* method using mammalian cells for the detection of cytotoxic effects of water samples is recommended in a guideline. However these *in vitro* methods are regulated in the DIN EN ISO 10993-5:2009-10 guideline for the biological evaluation of medical devices which includes the standard procedure for four different cytotoxicity methods (Table 5) [87].

Table 5: Cytotoxicity tests required according to DIN EN ISO 10993-5:2009-10

Test method	Toxicological endpoint
Neutral Red Test	Lysosomal activity
Colony formation Test	Ability to form colonies
MTT Test	Mitochondrial activity
XTT Test	Mitochondrial activity

However, despite these four tests, many others are available (e.g. trypanblue assay, LDH) to investigate the variety of possible cytotoxic effects.

2.2.2. Genotoxicity

The field of genotoxicity deals with the identification of DNA damages previous to mutations. These tests therefore give information on the level of structural DNA damages but do not necessarily provide evidence for mutations. A variety of test systems have been developed in order to detect changes at the DNA level. However

Introduction

these methods are subjected to steady changes due to new demands. Therefore guidelines have been established to set basic conditions to perform genotoxicity tests. However, as can be seen in Table 6 not every genotoxicity test is regulated by a guideline. In regard to the toxicological evaluation of waste waters the OSPAR convention (The Convention for the Protection of the Marine Environment of the North-East Atlantic) established an expert group on whole effluent assessment. In 2002 this group published a report summarizing and providing guidelines for the methods used to detect genotoxic effects in water bodies [88].

Table 6: Selection of *in vitro* tests for the detection of genotoxic effects

Test name	Mode of action/ endpoint	Organism/cell line	Guideline	Reference
Chromosomal aberration test	Structural chromosomal abberations	Mammalian cells	OECD TG 473 EPA-HQ-OPPT-2009-0156-0029	[89-91]
Sister Chromatid Exchange	Reciprocal DNA exchanges	Mammalian Cells	OECD TG 479	[92, 93]
Unscheduled DNA Synthesis	DNA repair and synthesis	Mammalian cells	OECD TG 482	[94]
Micronucleus Assay	Micronuleus formation	Mammalian cells	OECD TG 487	[95, 96]
Alkaline DNA-Elution assay	DNA strand breaks	Mammalian cells		[97-99]
p53 Calux test	Reporter gene Assay	HepG2/U2OS		[100]
Comet Assay (Single cell gel electrophoresis assay)	Single/double strand DNA breaks, alkali labile sites	Mammalian cells	Guidelines for *in vitro* and *in vivo* genetic toxicology testing OECD Guideline in progress	[101-103]
GreenScreen HC assay	DNA damage response/ stress pathway	p53-competent TK6 lymphoblastoid cell line		[104]
Toxicogenomics	Gene expression	Mammalian cells		[105-107]

Introduction

2.2.3. Mutagenicity

A mutation describes a DNA damage which can be transferred to the offspring, thus being hereditary. This is in contrast to general genotoxic effects, which are characterized as structural damages. Therefore mutagenic effects can be described as a subgroup of genotoxic effects [82]. Different kinds of mutations are known depending on their effect (negative, positive or no effect) or their range (gene, chromosme or genome) [82, 83]. Table 7 displays a selection of available *in vitro* test systems for the detection of mutagenic effects.

Table 7: Selection of *in vitro* tests for the detection of mutagenic effects

Test name	Mode of action/ endpoint	Organism/ cell line	Guideline	Reference
In vitro Mammalian Cell Gene Mutation Test (TK, HPRT, XPRT)	Gene mutations	Mammalian cells	OECD TG 476 EPA-HQ-OPPT-2009-0156	[82, 108, 109]
Mouse lymphoma test	Gene mutations	L5178Y	OECD 476	[108, 110]
Gene Mutation Assay	Base substitution, Frameshift	*Saccharomyces cerevisiae*	OECD TG 480	[111]
Mitotic Recombination Assay	Gene conversion, crossing over	*Saccharomyces cerevisiae,*	OECD TG 481	[112]
Ames Test	Base substitution, deletion, addition	*Salmonella typhimurium*	OECD TG 471	[113-115]
Umu Test	Base substitution, deletion, addition	*Escherichia coli*	OECD TG 471	[113]

2.2.4. Estrogenicity

The endocrine system enables the communication between different organs and parts of a body using hormones to regulate vegetative functions [116]. A great number of studies have been published showing the effects of substances with estrogenic activity on the ecosystem. Effects include influences on the reproduction, immune system, damages to sexual organs or the brain or influences on the embryonal development [71, 117-120]. In order to detect disturbances to this system through endocrine disruptors, tests with a low sensitivity are needed since hormones are able to induce effects at very small concentrations. For the testing of the estrogenic activity of a substance a variety of *in vitro* methods are available (Table 8).

Table 8: Selection of *in vitro* tests for the determination of estrogenic effects

Test name	Mode of action/endpoint	Organism/ cell line	Guideline	Reference
Yeast estrogen screen	Receptor binding assay (human ER)	*Saccharomyces cerevisiae*		[121]
E-screen	Estrogen dependent proliferation	MCF-7		[122, 123]
rt-YES	Receptor binding assay (fish ER)	*Saccharomyces cerevisiae*		[124, 125]
A-Yes	Receptor binding assay (human ER)	*Arxula adeninivorans*		[126, 127]
ER CALUX	Reporter gene assay	T47D/U2-OS		[128]
Ceri-hER-alpha	Transcriptional Activation	HeLa-9903	OECD TG 455	[129]
MVLN	Reporter gene assay	MVLN		[130, 131]
RTG-2 Assay	Reporter gene assay	RTG-2		[132]
H295R	Steroidogenesis Assay	NCI-H295R	OECD TG 456	[133, 134]
ELRA	Enzyme-Linked-Receptor-Assay	Purified estrogen receptor		[135]

Introduction

Due to the variety of methods available, *in vitro* and *in vivo*, the OECD initiated a task force on Endocrine Disruptor Testing and Assessment (EDTA) in the late 1990s [136]. In 2002 the Conceptual Framework for the Testing and Assessment of Endocrine Disrupting Chemicals was then released [137]. This framework includes five levels of testing starting with a classification/sorting based on available information (level 1) followed by *in vitro* assays (level 2) and *in vivo* testing (levels 3 to 5) suggesting test systems to be used. Kase et al. (2009) also supposed the use of a combination of *in vitro* and *in vivo* test systems to evaluate endocrine effects as well as others [71, 138]. This is in accordance to the OECD Conceptual Framework which supposes a simultaneous measurement of cytotoxic and estrogenic effects [137].

Although the number of available test systems for the evaluation of estrogenic effects is high, only a few of them are regulated through guidelines. Therefore in 2011 a DIN working group has been formed in Germany to establish a guideline for the estrogenicity testing in water systems (DIN-Arbeitskreis NA 119-01-03-05-09 AK "Hormonelle Wirkungen (Xenohormone)).

Despite the variety of available toxicological test systems only few of them are applied in water testing on a regular basis. Just a few of these methods are regulated through guidelines and are not part of any guideline for the evaluation of water quality.

3. Aims of the study

Micropollutants are commonly detected in the aqueous environment because they are not sufficiently removed durig conventional waste water treatment. These micropollutants include biocides, pharmaceuticals, personal care products as well as other chemicals. Since many of them are designed to be biologically active they might pose a risk to the environment. In addition for most of these substances knowledge about their environmental fate and toxicity is scarce. Thus additional waste water treatment steps are needed. Therefore the aims of this study were

- the adaptation and optimization of toxicological *in vitro* test methods for the testing of water samples
- the application of toxicological methods for the comparison of ozonation and UV oxidation with H_2O_2 in regard to the removal of organic micropollutants
- the comparison of ozonation and UV oxidation with H_2O_2 in regard to the formation of toxic oxidation by-products
- the toxicological evaluation of micropollutants and their oxidation by-products before and after oxidative treatment

4. Material and Methods

The chemicals and materials as well as the equipment used during this study are listed in the annex (chapter 8). The solutions used for the specific tests are listed below.

4.1. Solutions
4.1.1. Cell Culture
4.1.1.1. CHO cells

Table 9: Cell culture solutions for CHO cells

Solution	Composition
Growth Medium	HAM's F12 with 10 % FCS, 3 mL Gentamycin and 3 mL L-Glutamine
Freezing medium	HAM's F12 with 15 % FCS and 7.5 % DMSO
Trypsine (0.05 %)	Trypsine and EDTA (0.2 g/L)

4.1.1.2. T47D cells

Table 10: Cell culture solution for T47D cells

Solution	Composition
Growth Medium	DMEM F12 with 7.5 % FCS, 3 mL Gentamycin and 5 mL NEAA
Freezing medium	7 mL DMEM F12, 1 mL DMSO and 2 mL FCS
Trypsine (0.05 %)	Trypsine and EDTA (0.2 g/L)

Material and Methods

4.1.2. PAN I Multitox Test

Table 11: Solutions for the PAN I Multitox test

Solution	Composition
Solvent control	4.5 mL HAM's F12 and 0.5 mL Millipore water
Positive control	0.7 mL HAM's F12 and 1 % Triton® X-100
LDH II / LDH III mix (1 plate)	16 mL LDH II and 3.4 mL LDH III
XTT I / XTT II mix (1 plate)	4 mL XTT I and 40 µL XTT II

4.1.3. MTT Test

Table 12: Solutions for the MTT test

Solution	Composition
Solvent control	4.5 mL HAM's F12 and 0.5 mL Millipore water
Positive control	0.7 mL HAM's F12 and 1 % Triton® X-100
MTT solution	0.125 g 3-(4,5-Dimethylthiazol-2-yl)-2,5-diphenyltetrazoliumbromid dissolved in 25 mL DPBS and filter sterilized
Lysis solution	99.4 mL DMSO, 0.6 mL 100 % acetic acid and 10 g SDS

Material and Methods

4.1.4. Alkaline Comet Assay

Table 13: Solutions for the Alkaline Comet Assay

Solution	Composition
ENU stock solution (10 mg/L)	10 mg ENU dissolved in 1 L Ampuwa
Tris-Solution (1 M)	157.6 g Trizma dissolved in 1 L Ampuwa
NaOH Solution (2 M)	80 g NaOH dissolved in 1 L Ampuwa
EDTA (0.5 M)	186.1 g EDTA dissolved in 1 L Ampuwa and adjusted to pH 8 with NaOH (2M)
Lysis solution I	10 mL Tris-solution (1 M), 146.1 g NaCl, 100 mL EDTA (1 M) und 10 g N-Laurylsarcosine Sodium Salt were filled up to 1 L with Ampuwa and heated to 100 °C
Lysis solution II	100 mL DMSO and 10 mL Triton-X were mixed and stored protected from light
Neutralization solution	200 mL Tris-solution (1 M) were filled up to ~500 mL with Ampuwa and adjusted to pH 7.5 with NaOH (2M).
50X TAE Buffer	242 g Trizma, 57.1 mL glacial acetic acid and 100 mL EDTA (0.5 M) filled up to 1 L with Ampuwa
1X TAE Buffer	10 mL 50X TAE Puffer diluted with 490 mL Ampuwa
L.M.P. Agarose (0.75%)	0.75 g Low Melting Point Agarose dissolved in 100 mL PBS
Electrophoresis solution	75 mL NaOH (2 M), 1 mL EDTA (0.5 M) and 0.79 g Trizma filled up to 500 mL with Ampuwa and adjusted to pH 12.7 using HCl
SYBR®-Green stock solution	1 mg/mL SYBR®-Green in DMSO
SYBR-Green® staining solution	5 µL SYBR-Green® stock solution added to 50 mL 1X TAE Buffer

4.1.5. Ames Test

Table 14: Solutions for the Ames Test

Solution	Composition
S9 Mix	1.438 mL S9 buffer salts, 0.063 mL S9 G-6-P, 0.250 mL S9 NADP, 0.750 mL S9 Fraction

4.1.6. ER Calux

Table 15: Solutions for the ER Calux

Solution	Composition
Assay Medium I	DMEM F12 (without phenol red) with 5 % stripped FCS and 5 mL NEAA
Assay Medium II	9 mL Assay Medium I and 1 mL Ampuwa
Assay Medium III	11 mL Assay Medium I and 11 µL DMSO
Lysis reagent *(400 mL; pH 7.8)* [purchased from BDS]	33.03 g Glycylglycine, 36.97 g $MgSO_4$, 15.22 g EGTA, 4 mL Triton® X-100; 0.2 M NaOH to adjust pH
GlowMix *(1 L; pH 7.8)* [purchased from BDS]	3.58 g Tricine, 0.52 g $C_4H_2Mg_5O_{14}$, 0.60 g $MgSO_4$, 0.037 g EDTA, 5.14 g DTT, 0.21 g Co-enzyme A, 0.15 g Luciferine, 0.29 g ATP

Material and Methods

4.2. Cell Culture methods

4.2.1. CHO cells

CHO cells (Figure 14) are a permanent epithelian cell line derived from the ovaries of the Chinese Hamster *Cricetulus griseus* by Puck *et al.* in 1957 [139]. These are adherent cells cultivated in HAM's F12 medium supplemented with 10 % FCS, 0.5 % L-Glutamine and 0.5 % Gentamycin and grown at 37 °C with 5 % CO_2 and 95 % humidity.

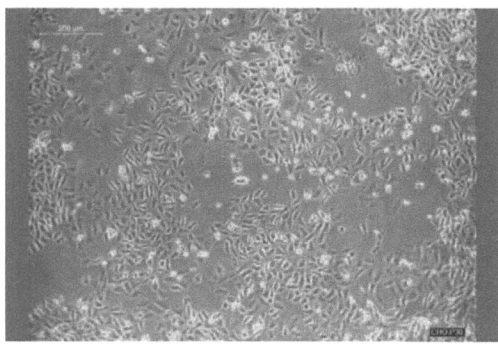

Figure 14: CHO cells; passage 30

4.2.1.1. Thawing

Before the cells are thawed a cell culture flask (25 cm^2) is filled with 10 mL preheated HAM's F12 medium (37 °C). The cryovial is then taken out of the liquid nitrogen and quickly thawed using a waterbath (37 °C) until only a small frozen part is left. Subsequently the cell suspension is transferred to the prepared culture flask and stored in the incubator (37 °C, 5 % CO_2, 95 % humidity). The next day the medium is changed and the cells are again incubated until further use.

4.2.1.2. Subculturing

For passaging the medium is removed and the cells are washed once with PBS. Culture flasks with a confluent cell growth are then trypsinated for 10 to 15 seconds using a 0.05 % trypsin solution with EDTA. After discarding the trypsin the cells are incubated for 3 minutes at 37 °C. The trypsinated cells are then resuspended in HAM's F12 medium and counted with a Neubauer Chamber to determine the total cell number. Depending on the further use, new culture flasks or well plates are then seeded with the appropriate cell number.

4.2.1.3. *Freezing*

For the cryoconservation the CHO cells are trypsinated as described above and resuspended in 15 mL HAM's F12 medium. The cell suspension is then transferred to a 15 mL vial and centrifuged at room temperature and 1200 rpm for 5 min. After centrifugation the supernatant is removed and the cells are resuspended in cooled (4 °C) freezing medium (HAM's F12 with 15 % FCS and 7.5 % DMSO) and aliquots of 1.8 mL are transferred to cryovials. The vials containing the cell suspension are then frozen in a three step process. First the vials are kept at -20 °C for 1 hour and then stored at -80 °C over night before they are kept in liquid nitrogen until further use.

4.2.2. T47D cells

T47D cells (Figure 15) are cultivated in DMEM F12 (Dulbecco's Modified Eagles Medium: Nutrient Mixture F12) medium with phenolred in a CO_2 incubator at 37 °C with 5 % CO_2 and 95 % humidity. These cells are a human breast adenocarcinoma cell line which is stably transfected with an estrogen-responsive luciferase reporter gene (pEREtata-Luc). This pEREtata-Luc reporter gene regulates the response of a luciferase reporter gene construct containing three human estrogen-responsive elements [128].

Material and Methods

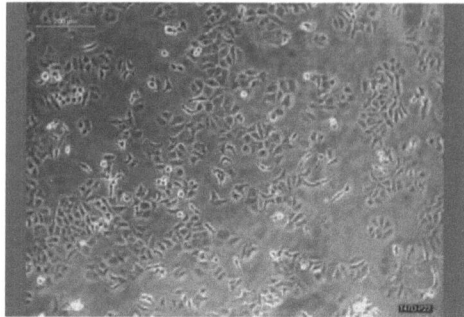

Figure 15: T47D cells; passage 22

4.2.2.1. Thawing

Three new culture flasks are filled with 10 mL of growth medium and kept in the incubator for 4 h. The cells in the cryovial are thawed until just a little ice is visible before 0.5 mL of growth medium from the culture flask is added to the cryovial. Then the cell suspension is transferred into the three previously prepared culture flasks as described below, before the flasks are stored in the incubator until further use:

Culture flask 1: 150 µL cell suspension
Culture flask 2: 450 µL cell suspension
Culture flask 3: 900 µL cell suspension

4.2.2.2.   Subculturing

When a confluent growth of 85 – 95 % is reached the cells are subcultured into new culture flasks. For this the medium is removed, the cells are washed with 5 mL PBS and trypsinated with 2 mL trypsin for 10 seconds. The trypsin is then removed and the flask is stored in the incubator for 5 minutes. After this step, the cells are resuspended in 10 mL growth medium and the appropriate volume of suspension is added to a new culture flask.

4.2.2.3. Freezing

The freezing medium is prepared fresh and stored on ice until use. The cells are trypsinated as described above and resuspended in 10 mL growth medium. The

Material and Methods

suspension is transferred into a sterile tube and centrifuged at 1100 rpm for 5 minutes. Then the supernatant is removed and the cell pellet is resuspended in 4 mL of ice-cold freezing medium. 1 mL of this suspension is then pipetted into the cryovials which have been stored on ice. After storing the cryovials at -80 °C for one day they are transferred to liquid nitrogen for long-term storage.

4.3. Toxicological methods

4.3.1. Cytotoxicity

4.3.1.1. PAN I: LDHe – XTT – NR – SRB

For the determination of cytotoxicity the PAN I: LDHe-XTT-NR-SRB test was used. This test allows the combined colorimetric detection of plasma membrane integrity, mitochondrial activity, lysosomal integrity and the rate of protein synthesis of cells exposed to various substances, i.e. chemicals, pharmaceuticals or other anthropogenic compounds. The advantage of this approach is the sequential measurement of four parameters using just one cell culture.

LDHe

This test allows the detection of the extracellular lactate dehydrogenase by measuring the oxidation rate of NADH to NAD^+ as well as the reduction of pyruvate to lactate:

$$Pyruvate + NADH + H^+ \rightarrow Lactate + NAD^+$$

The lactate dehydrogenase (LDH) is a stable enzyme present in the cytoplasm of all cells. Damages to the cell membrane result in a release of LDH into the cell culture medium. However since LDH has a molecular weight of 140 kDa it does not easily penetrate the cell membrane and a high amount of damage is needed in order to

Material and Methods
___

detect LDH outside the cytoplasm [140]. This release can then be determined kinetically by measuring the consumption of NADH.

XTT

The mitochondrial activity is determined using the XTT test. XTT (2.3-bis(2-methoxy-4-nitro-5-sulfophenyl)-2H-tetrazolium-5-carboxyanilide inner salt) is a yellow tetrazolium salt which gets cleaved to an orange formazan by the enzyme succinate dehydrogenase (Figure 16) [140]. For their survival viable cells need an intact mitochondrial respiratory chain. In the case of an exposure to a cytotoxic substance which affects the mitochondria, XTT will not be cleaved and a change in enzyme activity can then be measured photometrically. The reduction rate of XTT therefore directly correlates with the mitochondrial activity/integrity.

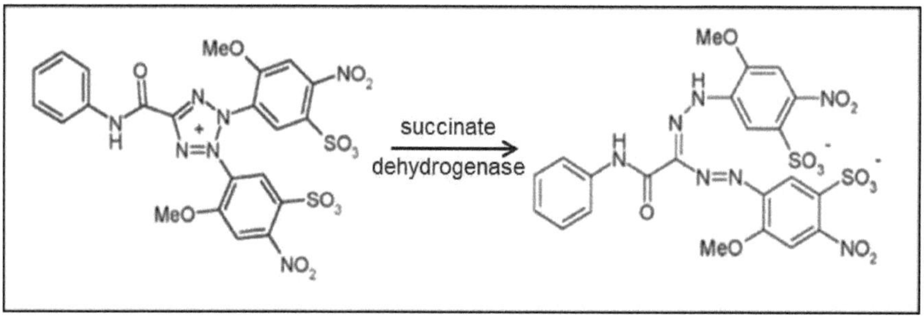

Figure 16: Chemical equation of the XTT cleavage to XTT formazan by succinate dehydrogenases

Neutral Red (NR)

Neutral Red (Figure 17) is a cationic dye used to determine the viability of cells since it penetrates the cell membrane by nonionic passive diffusion and binds to anionic and phosphate groups within the lysosomes. This is possible due to the pH gradient

Material and Methods

between the cytoplasm and the lysosomes (lysosomal pH (4.5 - 5) < cytoplasmic pH (~7)) which results in a charging of the dye once it reaches the lysosomes. Thus the dye is kept inside the lysosomes [141]. Lysosomes are cellular structures which contain enzymes to break down waste and cellular debris [142]. Effects on the cell surface or the lysosomal membrane result in a decreased incorporation and binding of neutral red. The amount of dye taken up by the cells can then be measured photometrically and is proportional to the number of viable cells and allows a differentiation between viable, damaged and dead cells.

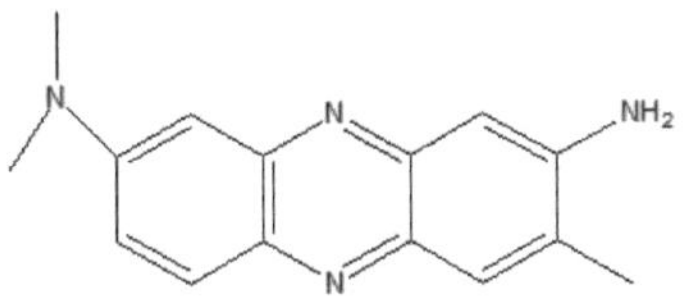

Figure 17: Structural formula of neutral red

Sulforhodamine B (SRB)

Sulforhodamine B (Figure 18) is an aminoxanthene-dye used for the detection of cytotoxic effects. The anionic dye binds electrostatically to cellular proteins thus being a toxicological marker for cell proliferation or cell death [143]. The number of viable cells can therefore be directly correlated with the total protein synthesis.

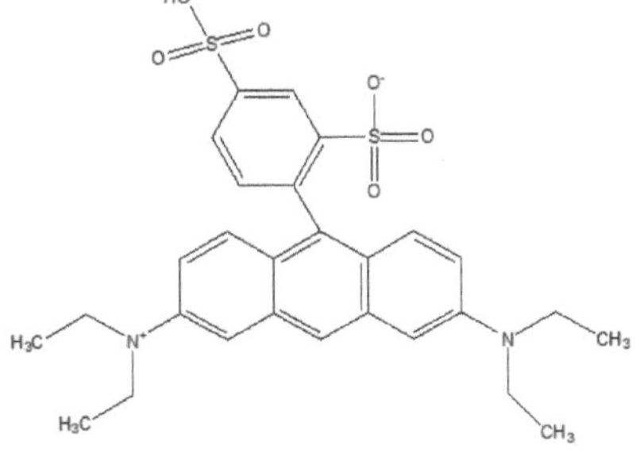

Figure 18: Structural formula of Sulforhodamine B

Material and Methods

4.3.1.1.1. Procedure

Exposure

CHO cells are trypsinated and resuspended as described above. The cell number is adjusted to 10,000 cells/200 µL medium. 200 µL of sterile PBS are added to the outer wells of a 96-well microtiter plate. Then 200 µL of the adjusted cell suspension is transferred to the 60 inner wells of the plate. The plate is then stored in the incubator for 24 h. Exposure of the cells to 200 µL of the controls is then done as illustrated below (Figure 19). The cells are exposed to the samples by adding 180 µL of HAM's F12 and 20 µL of the sample. After this exposure step the plates are then stored in the incubator for 24 h before further analysis.

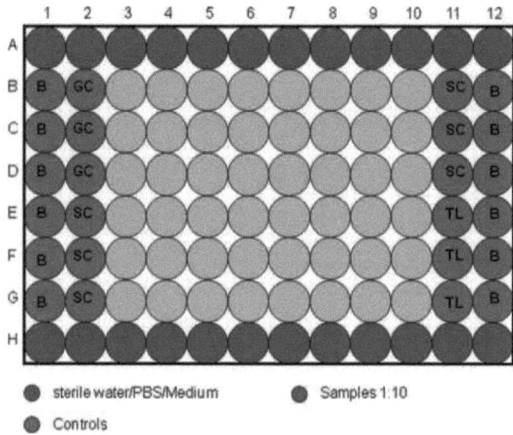

Figure 19: Layout of the 96 well plate seeded with CHO cells (20000 cells/0.2 mL). B = 200 µL blank control: culture medium + solvent without cells; GC = 200 µL cell growth control: culture medium + cells; SC = 200 µL solvent control: culture medium + cells + solvent; TL = 200 µL positive control: culture medium + cells + 1 % Triton® X-100

Cell handling

After 24 h of exposure 20 µL of the supernatant from each exposed well is transferred to a new 96 well plate and the plate containing the cells is again incubated until further use. 20 µL of LDH II/LDH III mix is subsequently added to each well containing the supernatant and the kinetic reading is started immediately. After this the original plate containing the cells is taken from the incubator and the cells are washed once with 200 µL PBS before 200 µL of fresh culture medium is

Material and Methods

added into each well. 50 µL/well of freshly prepared XTT I/XTT II mix is added and the plate is then further incubated for 2 h at 37 °C and 5 % CO_2. After incubation the contents of each well are carefully mixed and the OD_{480} is measured.

For measuring the Neutral Red (NR) uptake the XTT staining solution is removed and each well is washed with 300 µL of NR I solution. Then 200 µL of NR II labeling solution are pipetted into each well and the plate is again incubated for 3 h. The NR II solution is then removed and 200 µL of NR III fixing solution are added and discarded after 1 min. Finally 200 µL NR IV solubilization solution are added to each well, the plate is incubated at room temperature for 15 min, gently mixed and the OD_{540} is finally determined. The last step of the cytotoxicity detection is the SRB test. Therefore the NR IV solution is removed and the cells are washed with 300 µL SRB I washing solution followed by the addition of 250 µL SRB II fixing solution and 1 h of incubation. Then the cells are washed three times with 200 µL deionized water and 50 µL SRB III labeling solution are added. The plate is then again incubated for 15 min at room temperature and washed two times with 400 µL SRB IV rinsing solution. After this washing step the plates are air dried overnight. Then 200 µL of SRB V solubilization solution are added to each well and the plate is further incubated for 30 min at room temperature. The OD_{540} is then measured after gently mixing.

4.3.1.1.2. Measurement

The OD was measured with the TECAN GENios platereader (TECAN; Crailsheim, Germany)

LDHe: absorbance at 340 nm for 25 minutes measuring kinetically every 5 minutes

XTT: OD at 480 nm

NR: OD at 540 nm

SRB: OD at 540 nm

Material and Methods

4.3.1.1.3. Data analysis

LDHe: First the OD_{340} values from the kinetic reading are plotted vs. time to visualize the activity of the tested samples. From this graph a time interval is chosen where the curves are almost linear and the ΔOD/min for each well is calculated and the extracellular amount of NADH is presented as "nm consumed NADH/min/mL" (Figure 20).

$$\text{NADH consumed} = \frac{\Delta OD/min \times 0.260 \times 1000}{6.2 \times 20}$$

Figure 20: Equation used to calculate the extracellular NADH amount

With:
0.260 = reaction volume [mL]
1000 = gives the result in mL
6.2 = NADH extinction coefficient at 340 nm [mM]
20 = volume used in the assay [µL]

Based on the calculation of the amount of extracellular LDHe the viability of the cells is finally calculated (Figure 21).

$$\text{Viability [\%]} = 100 * \frac{\text{mean OD treated cells}}{\text{mean OD growth control}}$$

Figure 21: Equation used to calculate the percent of viable cells

An overview of the steps needed to finally calculate the number of viable cells using the LDHe test is presented in Table 16.

Material and Methods

Table 16: Steps of calculating the viability of the LDHe test

Step	Calculation
1.	$t_{start\ linearity} - t_{end\ linearity}$
2.	$(t_{start\ linearity} - t_{end\ linearity})/min$
3.	$(t_{start\ linearity} - t_{end\ linearity})/min$ - blank
4.	Units NADH consumed
5.	Viability

The % viability in the XTT, the NR and SRB test is calculated the same way. First the mean OD values for each sample and control are calculated. Then the mean OD values of the samples and the controls are corrected by subtracting the mean OD of the blank. Finally the viability is determined using the following equation:

$$\text{Viability [\%]} = \frac{100 * \text{corrected mean OD sample}}{\text{corrected mean OD solvent control}}$$

4.3.1.2. MTT Test

The MTT test is a colorimetric assay used to detect cytotoxic effects based on the cleavage of the MTT tetrazolium salt (3-(4,5-dimethylthiazol-2-yl)-2,5-diphenyl tetrazolium bromide). The principle of the MTT test is the same as for the XTT test with the difference in the solubility of the tetrazolium salts. Mitochondrial dehydrogenases cleave the MTT resulting in an unsoluble formazan (Figure 22) thus measuring the mitochondrial activity. Cells which are exposed to a cytotoxic sample might have a compromised mitochondrial respiratory chain resulting in a lesser content of the formazan which allows a direct correlation of transformed MTT and the number of viable cells [144, 145].

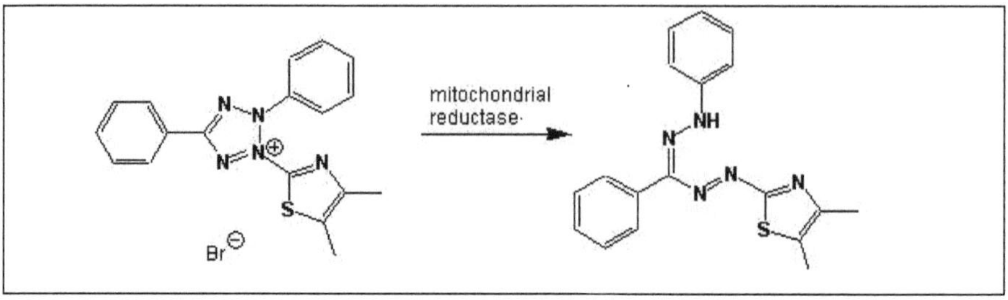

Material and Methods

Figure 22: Chemical equation of the MTT cleavage to MTT formazan

4.3.1.2.1. Procedure

Exposure

CHO or T47D cells were trypsinated and seeded into a 96 well plate with 20,000 cells/200 µL into each of the inner 60 wells. The outer wells were filled with 200 µL PBS and the plate was stored in the incubator (37 °C, 5 % CO_2, 95 % humidity) for 24 h. Before exposure the medium was removed and 180 µL fresh medium as well as 20 µL of the sample were added to the appropriate well. 50 µM MMA III were used as a positive control. The cells were then further incubated for another 24 h. Each sample and control was tested in three individual experiments per test.

Cell handling

After 24 h of incubation the exposure medium was removed, 100 µL fresh medium and 10 µL MTT solution were added, the plate was gently shaken for 5 min at room temperature and then again incubated for 2 h before the supernatant was removed and 100 µL lysis solution were added. After 5 min the plate was shaken for five more minutes before the OD was measured.

Material and Methods

4.3.1.2.2. Measurement

After lysing and incubating the cells the OD was measured at 595 nm and a reference wavelength of 620 nm using the TECAN GENios (TECAN; Crailsheim, Germany) microplate reader.

4.3.1.2.3. Data analysis

For each single experiment the viability of the cells was determined by calculating the mean percent of viable cells compared to the negative control using the following equation:

$$\text{Viability [\%]} = 100 * \frac{\text{mean OD treated cells}}{\text{mean OD negative control}}$$

The mean and the standard deviation of three individual experiments are then displayed graphically.

4.3.2. Degree of Cytotoxicity

The calculated viability resulting both from the MTT test and the PAN I test was then classified according to the DIN EN ISO 10993-5 guideline [146] as displayed in Table 17.

Table 17: Degree of cytotoxicity

Degree of damage	Viability [%]	Cytotoxicity
0	81 – 100 %	no cytotoxicity
1	71 – 80 %	weak cytotoxicity
2	61 – 70 %	moderate cytotoxicity
3	0 – 60 %	strong cytotoxicity

4.3.3. Genotoxicity

4.3.3.1. Alkaline Comet Assay

The Comet Assay is a technique used to measure DNA damage. These damages can result from two different types of effects. Either endogenous e.g. through reactive oxygen species (ROS) formed during metabolic activity, or exogenous caused by different types of radiation or chemicals, thus substances from outside the cell. DNA damaging agents can then induce a variety of effects on the DNA. One type of damage is the DNA strand break. These breaks can occur either directly by ROS attacking the sugar-phosphate backbone of the DNA or through the activity of endonucleases during DNA repair processes. Another type of DNA damage are alkali-labile sites. These are then transformed into DNA single-strand breaks under alkaline conditions and can therefore also be detected using the alkaline comet assay [82].

The Comet Assay is a sensitive microgel electrophoresis technique to detect DNA damage in single cells and was first described by Ostling et al. in 1984 [102]. They embedded the cells in agarose, lysed them in a neutral detergent solution and applied a weak electric field. During electrophoresis the positively charged DNA migrates towards the anode. In case of a damage DNA fragments of different sizes will migrate in a certain pattern which looks like a comet, while undamaged DNA appears as a solid circle (Figure 23). Various fluorescent stains which bind DNA can then be used to visualize the DNA and to analyze its degree of damage by means of the migration pattern. However, the limitation of this approach is the neutral lysing condition since this only allows the detection of double strand breaks. Therefore Singh et al. (1988) further optimized this method. They used alkaline conditions during lysis and electrophoresis which resulted in a more sensitive assay detecting both single strand and double strand DNA breaks as well as alkali labile sites [103]. In the present study I further modified the method described above for the use of CHO cells and the testing of water samples according to the procedures described by Singh et al. and to the guidelines for *in vitro* testing published by Tice et al. (2000) [101].

Material and Methods

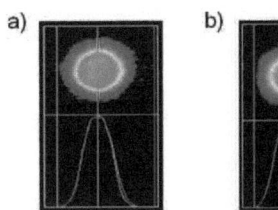

Figure 23: Comet Assay analysis of undamaged (a) and damaged (b) CHO cells stained with SYBR Green® and analyzed using the Comet Assay 4 Software.

4.3.3.1.1. Procedure

Exposure:

CHO cells were cultured in HAM's F12 medium. 24 h prior to exposure the cells were seeded in 24 well plates with 100,000 cells per well in 2 mL medium. Before exposure the medium was removed and fresh medium was added. The cells were then exposed to the samples for 24 h at a 1:10 ratio (1.8 mL medium and 0.2 ml sample). 0.1 mg/mL N-ethyl-N-nitrosourea (ENU) were used as a positive control and the cells were exposed for 30 min.

Microgel preparation:

Microgels were prepared by sticking a chamber slide with eight chambers to a GelBond® film. Each chamber was sealed by adding 50 µL low melting point (L.M.P.) agarose which was then allowed to solidify on ice.

Cell handling:

After exposure the medium was discarded and the cells were washed with 1 mL PBS. Then the cells were trypsinated for 10 sec followed by 3 min of incubation at 37 °C without trypsin. Depending on their confluency they were then resuspended in PBS, counted, and the cell number was adjusted to 8000 cells/20 µL. These 20 µL of each cell suspension were then mixed with 45 µL L.M.P. agarose and added to the prepared chambers. After all gels were solidified, the chamber slides were removed and the film was placed in a 4 °C tempered lysis solution and stored at 4 °C over night.

Electrophoresis:
For electrophoresis the freshly prepared solution was cooled down to 4 °C before use. Prior to applying the electric field the films were incubated in electrophoresis solution for 20 min and the electrophoresis was then run for 20 min at 300 mA. Then the films were transferred to neutralization solution for 30 min and further transferred to ethanol absolute for 2 h before the gels were left to dry over night.

Staining:
The dried gels were stained with the SYBR-Green® working solution for 18 min and gently washed with water. To avoid air bubbles a few water drops were added onto the film before a cover glass was applied.

4.3.3.1.2. Analysis

For analysis the Comet Assay 4 software from Perceptive Instruments was used. Each gelbond film contains eight microgels including a positive and a negative control (Figure 24a). The gels were analyzed in an s-shaped pattern (Figure 24b) always starting at the top left corner until 50 nuclei had been scored. As a measure of DNA damage the Olive Tail Moment [147] is displayed graphically

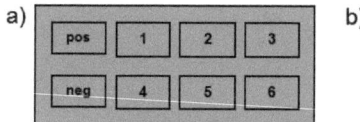
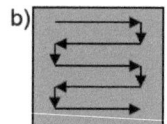

Figure 24: a) Gelbond film containing eight microgels. b) Scoring pattern of a single microgel. pos = positive control; neg = negative control; 1 – 6 = gels containing cells exposed to the samples.

Material and Methods

4.3.3.1.3. Statistics

Statistical analysis was done using GraphPad Prism column analysis. The data of three individual experiments were summarized and plotted using their mean value and the standard error of mean. The significance of DNA damage compared to the negative control was analyzed using the Mann-Whitney Test and classified according to the p-value (Table 18).

Table 18: Significance of DNA damage according to the Mann Whitney Test

P-value	> 0.05	0.01 – 0.05	0.001 – 0.01	< 0.001
P-value summary	n.s.	*	**	***
Significance	not significant	significant	very significant	highly significant

Material and Methods

4.3.4. Mutagenicity

4.3.4.1. Ames Test

Mutagenic activity of water samples containing chemicals and other substances was detected using the Ames MPF™ 98/100 Aqua. The two used histidin-dependent *Salmonella typhimurium* strains TA98 and TA100 either carry a frameshift mutation (TA98) or a base substitution mutation (TA100) in their histidin operon [148]. Thus they are not able to synthesize histidin themselves and need histidin-containing medium to grow. Once the *Salmonella* are exposed to samples with mutagenic activity and a frameshift or base substitution mutation takes place the bacteria are able to grow in histidin-free medium, since they regained their ability to synthesize histidin [149].

The test is then performed in two different ways. Due to the fact that some substances are only activated through metabolization and bacteria lack the appropriate eukaryotic metabolic enzymes the exposure is done with and without the addition of a liver enzyme mix (S9 Mix). This addition results in an *in vitro* metabolization of the tested substances thus the mutagenic activity of metabolites can also be detected [148, 149].

4.3.4.1.1. Procedure

The Ames test was performed using the AMES MPF™ 98/100 Aqua version provided by Xenomtrix (CH). This test includes the *Salmonella typhimurium* strains TA98 and TA100. Two Erlenmeyer flasks were filled with 10 mL growth medium and 10 µL ampicillin (50 mg/mL). In addition a third Erlenmeyer flask was filled with 10 mL medium only as a negative control. The vials containing the bacterial suspension were thawed and 50 µL of the suspension were transferred to the prepared Erlenmeyer flasks which were then incubated for 16 h in a shaking water bath at 37 °C. After incubation the OD was measured at 600 nm. For that purpose 900 µL of growth medium and 100 µL of the bacterial suspension as well as the negative control were filled into cuvettes and measured. The samples were then tested in triplicate with and without metabolic activation by S9 using a 24-Well plate.

Material and Methods

For exposure the medium, samples, bacteria, water and S9 were mixed as shown in Table 19 and incubated at 37 °C for 90 min.

Table 19: Scheme of pipetting for exposure

	TA98		TA100	
	-S9	+S9	-S9	+S9
10x Exposure medium	25 µL	25 µL	25 µL	25 µL
Sample	185 µL	185 µL	185 µL	185 µL
Water	15 µL	--	15 µL	--
Bacteria	25 µL	25 µL	25 µL	25 µL
S9 Mix	--	15 µL	--	15 µL
Total volume	250 µL	250 µL	250 µL	250 µL

After incubation 2.8 mL of indicator medium were added to each well and mixed properly. Then the contents of each well were transferred to 48 wells of a 384-well plate by pipetting 50 µL into each of the 48 wells. After this the plates were placed into a plastic bag and then further incubated for 48 h at 37 °C.

4.3.4.1.2. Analysis

After incubation the plates were removed from the incubator and the number of positive wells containing revertant bacteria (wells where the medium has turned yellow) for each sample was counted (Figure 25). The results were then analyzed using an Excel data sheet provided by Xenometrix (Allschwil, CH).

Material and Methods

Figure 25: 384-Well plate after 48 h of exposure. The medium of the positive wells containing bacteria with reversed mutations has changed to yellow and the medium of the wells without mutated bacteria is purple.

4.3.4.1.3. Statistics

The statistical analysis was done using the provided Excel data sheets (Xenometrix; Allschwil, CH). A sample was classified mutagenic when the amount of positive wells was two-times higher than the negative control (fold increase over baseline ≥ 2). Statistical analysis was done using the 1-sided t-test based on unpaired data.

The results are declared reliable when the quality standards listed in Table 20 are met. The number of positive wells thereby describes the wells which have turned yellow or contain a colony, thus wells containing bacteria with reversed mutations.

Table 20: Ames MPF® 98/100 quality standards

		Mean number of positive wells	
		TA98	*TA100*
Negative control	Solvent	≤ 8	≤ 12
Positive controls	4-NQO	≥ 25	≥ 25
	2-NF	≥ 25	≥ 25
	2-AA (only +S9)	≥ 25	≥ 25

4.3.5. Estrogenicity

4.3.5.1. ER Calux

The ER Calux (**E**strogen **R**eceptor - **C**hemical **A**ctivated **Lu**ciferase gene e**X**pression) is a test developed to detect estrogenic effects of chemicals and other substances. Estrogens, e.g. estradiol, might affect the thyroid function, the reproductivity, the nervous system as well as the cardiac system [150]. Thus increased amounts of these substances might pose a risk to human health. The test uses a modified T47D cell line which is a human breast adenocarcinoma cell line. These cells have been genetically modified to emit light with an intensity correlated to the amount of endocrine substances present in the sample. Therefore an estrogen responsive element (ERE) was coupled to a luciferase gene. Once the chemical reporter binds to the ERE it induces the expression of the luciferase gene associated with the ERE. The activity of the luciferase can then be detected and correlated to the amount of estrogenic substances in the sample [128].

4.3.5.1.1. Procedure

Exposure

48 h prior to incubation the 60 inner wells of a 96 well plate were filled with 100 µL cell suspension whereas the outer wells were filled with 200 µL PBS. For this culture flasks with 85 – 95 % confluent growth were used. The growth medium was removed and the cells were washed with 5 mL PBS before trypsination with 2 mL trypsin for 10 seconds. The flasks were then stored in the incubator for 5 minutes. Then the cells were resuspended in 10 mL assay medium I, counted and diluted with assay medium I to achieve a concentration of 100,000 cells/mL. After incubating the plates for 24 h at 37 °C and 5 % CO_2 the medium was removed and new assay medium I added, before the plates were again stored in the incubator for 24 h. To expose the cells to the 17ß-ethinylestradiol standards 1 µL of each standard was mixed with 1 mL of assay medium II. 100 µL of each solution were then added to the appropriate wells. The DMSO control was also prepared by mixing 1 µL DMSO with 1 mL assay medium II and the same was done for the water control. The samples were then added to the wells using a 1:10 dilution into assay medium III with a final

Material and Methods

volume of 100 µL/well and the plates (Figure 26) were incubated for 24 h before the analysis was performed.

	1	2	3	4	5	6	7	8	9	10	11	12
A												
B		E2-0	E2-0.3	E2-0.6	E2-1.0	E2-3.0	E2-6.0	E2-10	E2-30	DMSO	medium	
C		E2-0	E2-0.3	E2-0.6	E2-1.0	E2-3.0	E2-6.0	E2-10	E2-30	DMSO	medium	
D		E2-0	E2-0.3	E2-0.6	E2-1.0	E2-3.0	E2-6.0	E2-10	E2-30	DMSO	medium	
E		1	2	3	4	5	6	7	8	9	10	
F		1	2	3	4	5	6	7	8	9	10	
G		1	2	3	4	5	6	7	8	9	10	
H												

Figure 26: ER Calux exposure plate (BDS). E2 = 17ß-ethinylestradiol; E2-0 to E2-30 = 17ß-ethinylestradiol concentration [pM/well]; DMSO = wells containing 1 % DMSO; medium = wells containing medium only; 1 – 10 = wells containing the cells exposed to different samples.

Cell handling

After an exposure time of 24 h the medium was removed and 50 µL of lysis reagent was added to the exposed wells before the plates were stored at room temperature for 15 minutes. 30 µL of each well were then transferred to a new white 96-well plate.

4.3.5.1.2. Analysis

30 µL of GlowMix solution are then added to the 30 µL of supernatant and the luminescence was immediately measured using the TECAN GENios microplate reader resulting in an excel sheet displaying the measured values as relative light units. These values were then copied into an excel sheet provided by BDS allowing the automated analysis and determination of estradiol equivalents (EEQ) displayed as the mean pM EEQ/well. EEQ values were considered right when the following quality standards (Table 21) were met.

Table 21: ER Calux quality standards

Automatic verification	Cell on excel sheet
General parameter	
EC_{50} should be between 2 and 8 pM/well	D4
Induction should be ≥ 6	D5
R^2 should be ≥ 0.98	D8
Sample parameters	
Reaction of cells should be < EC_{50}	H30 – H40
Reaction of cells should be > LOQ (1.5 pM)	I30 – I40
Standard deviation should not be > 15 %	J30 – J40
No remarks were made	K30 – K40
Overall verification	
All above mentioned parameters should be answered with „true"	D15 – D25
Non-automatic verification	
The negative control (E2-0) should not be higher than the LOQ	
Verification of the reference materials	

Material and Methods

4.4. Oxidation of water samples

The water samples were prepared, oxidized and chemically analyzed either by the working group of Dr. Jochen Türk at the Institut für Energie und Umwelttechnik e.V. in Duisburg (IUTA e.V.) or by the working group of Dr. Kai Bester at the Department of Environmental Chemistry at the University Duisburg-Essen (UDE). Experimental conditions as well as the operating procedures were adapted from the final report of the IGF project No. 15862 N "Oxidationsnebenprodukte" [151]. The investigated substances and chemicals used for the oxidation experiments and chemical analyses are listed in Table 22.

Table 22: Chemicals and investigated substances used for oxidation experiments

Chemical	Company
2,4-Dichlorophenol	Sigma-Aldrich; Steinheim, D
Acetone	Merck; Darmstadt, D
Acetonitrile	LGC Promochem; Wesel, D
AHTN	Dr. Ehrensdorfer; Augsburg, D
Atenolol	Sigma-Aldrich; Steinheim, D
Bisphenol A	Dr. Ehrensdorfer; Augsburg, D
Catalase (*Aspergillus niger*)	Sigma-Aldrich; Steinheim, D
Ciprofloxacin	Sigma-Aldrich; Steinheim, D
Ethinylestradiol	Riedel-de Häen; Seelze, D
HHCB	Dr. Ehrensdorfer; Augsburg, D
HHCB-Lacton	Dr. Ehrensdorfer; Augsburg, D
HPLC-water	J.T.Baker; Deventer, NL
Hydrogen peroxide (H_2O_2)	Sigma-Aldrich; Steinheim, D
Irgarol 1051	Hempel; Pinnberg, D
Methanol	Merck; Darmstadt, D
Methyl tert-butyl ether (MTBE)	Merck; Darmstadt, D
Metoprolol	Sigma-Aldrich; Steinheim, D
N-Butyl Methyl Ether	J.T.Baker; Deventer, NL
Ofloxacin	Sigma-Aldrich; Steinheim, D
Sulfamethoxazole	Sigma-Aldrich; Steinheim, D
Terbutryn	Dr. Ehrensdorfer; Augsburg, D
Toluol	Merck; Darmstadt, D
Triclosan	Dr. Ehrensdorfer; Augsburg, D
Triphenylphosphate (TPP)	Sigma-Aldrich; Steinheim, D
Tris-(2-butoxyethyl)-phosphate (TBEP)	Sigma-Aldrich; Steinheim, D
Tris-(1-chloro-2-propyl)-phosphate (TCPP)	Akzo Nobel; Amersfoort, NL

Material and Methods

Stock solutions:

The substances were dissolved in HPLC-water containing 50 % acetonitrile with a concentration of 1 g/L. Substance concentrations during the oxidation experiments were between 0.1 and 18 mg/L. Table 23 displays the systems used for the oxidation of the individual samples.

Table 23: Systems used for oxidation experiments of the tested samples, and their final concentration during toxicological testing. IUTA = Institut für Energie- und Umwelttechnik e.V.; UDE = Institute of Environmental Chemistry, University Duisburg-Essen

Substance	System (Place)	Final concentration after exposure for toxicological tests
AHTN	(UDE)	0.1 mg/L
Atenolol	(UDE)	0.2 mg/L
Bisphenol A	(IUTA)	1.4 mg/L (Ozonation); 0.1 mg/L (UV/H_2O_2)
Ciprofloxacin	(IUTA)	1.4 mg/L (Ozonation); 0.1 mg/L (UV/H_2O_2)
Ethinylestradiol	(IUTA)	1.5 mg/L
HHCB	(UDE)	0 – 50 µg/L; 0.1 mg/L (UV/H_2O_2)
HHCB-Lacton	(UDE)	0 – 50 µg/L
Irgarol 1051	(UDE)	0.75 mg/L
Metoprolol	(IUTA)	1.4 mg/L (Ozonation); 0.1 mg/L (UV/H_2O_2)
Ofloxacin	(IUTA)	18 mg/L
Sulfamethoxazole	(IUTA)	1.4 mg/L (Ozonation); 0.1 mg/L (UV/H_2O_2)
TCEP	(UDE)	0.1 mg/L
TCPP	(UDE)	0.1 mg/L
TPP	(UDE)	0.1 mg/L
Terbutryn	(UDE)	0.49 mg/L
Triclosan	(UDE)	0 – 100 µg/L
2.4-Dichlorophenol	(UDE)	0 – 100 µg/L

Material and Methods

4.4.1. Ozonation

4.4.1.1. Laboratory scale ozonation

Institut für Energie und Umwelttechnik e.V. (IUTA e.V.):

During laboratory scale ozonation experiments (up to 1 L of sample), the ozone gas was produced using technical oxygen and an ozone generator (COM AD-01; Anseros, Tübingen, D). To produce ozone water, the gas was directed through distilled, cooled water immediately after generation. A platinum catalyzer was used to destroy the remaining excessive ozone. The ozone concentration was determined photometrically at 260 nm (SPECORD® PC 200 UV VIS Spectrophotometer; Analytik Jena AG, Jena, D) in regard to the Lambert-Beer law using a molar absorption coefficient of 3300 $mol^{-1}\ cm^{-1}$.

Pure water as well as WWTP effluents were then spiked with the single substances before the previously prepared ozone water was then added to the samples. After the reactions of the ozone with the water contaminants the samples were filter sterilized (Chromafil RC, 0.45 µm; Macherey-Nagel, Düren, D) and either directly stored in HPLC-MS-vials or after a subsequent solid phase extraction (SPE; see chapter 4.5.1). In addition WWTP effluent was also oxidized and analyzed without the addition of substances.

Department of Environmental Chemistry:

All tested substances were dissolved either in HPLC-water or WWTP effluent with concentrations between 0.1 and 44 mg/L. These solutions were then used for ozonation experiments. Residue-free methanol, toluol, acetone or MTBE were used to terminate the OH° formation. Mixing the solutions with O_3 at different ratios, samples with different mole-ratios were gained (1:1 up to 1:10). Ozone was produced using an ozone generator (Enaly 1000BT-12, Enaly M&E Ltd, Shanghai, China) and technical oxygen (0.5 L/min) resulting in an ozone input of 2-5 mg/L into the system. The amount of ozone was detected using UV-VIS spectrometry (Shimadzu, Duisburg, Germany) and samples were taken time-dependent.

Material and Methods

4.4.1.2. Pilot scale ozonation (IUTA e.V.)

To perform experiments with a sample volume of 200 to 500 L a pilot plant (OCS H_2O_2-UV; Wedeco, Herford, D) with an attached ozone generator (OCS-GSO 20; Wedeco, Herford, D) and an ozone surveillance device (DH 5; BMT Messtechnik. Berlin, D) was used. The ozone generator was adjusted to discharge 80 – 90 g/m^3 gaseous ozone into the pilot plant with a flow rate of 0.5 m^3/h. All experiments were performed at 20 °C and before starting an experiment the plant was rinsed with the used water. Samples were then taken in a time-dependent manner.

4.4.2. UV/H_2O_2 oxidation

4.4.2.1. Laboratory scale UV oxidation with and without H_2O_2 (IUTA e.V.)

Oxidation experiments using UV light were performed at a laboratory scale system with a volume of 1 L. The construction was coupled to a low pressure UV lamp (TNN15/32 Hg lamp, 254 nm; Heraeus, Hanau. D). Before the oxidation, spiked and non-spiked pure water as well as WWTP effluents were evenly distributed in the system and the oxidation was started. At the beginning of the experiments performed with the addition of H_2O_2 the UV lamp was allowed to reach full power (5 min) before H_2O_2 (0.3 g/L) was added. To remove remaining peroxides after treatment catalase (0.25 mL; c = 0.4 kU/mL) was added to the system. All experiments were performed at 30 °C and sampling was done time-dependent.

4.4.2.2. UV/H_2O_2 oxidation of TPP, TBEP and TCPP (UDE)

Each of the three organophosphates was dissolved in 1 L HPLC-water with a final concentration of 1 mg/L and subjected to UV treatment. The samples were treated with UV light (Heraeus TNN15/32; 3 Watt, 254 nm; Hg-LP; Heraeus, Hanau, Deutschland) and 1 mg/L H_2O_2 for up to 120 min with a time-dependent sampling. After sampling, 100 µL catalase (0.4 kU/mL) were added to remove remaining peroxides. Subsequent a SPE (see chapter 4.5.2) was performed before GC-MS-Screening.

4.4.2.3. Pilot scale UV oxidation with and without H_2O_2 (IUTA e.V.)

The previously described pilot scale plant (Volume = 200 L) was also used for UV experiments. All samples were UV treated the same way as was done during ozonation. For oxidation a low pressure UV lamp (XLR 10, 33 W, 180 nm and 254 nm; Wedeco, Herford, D) or a medium pressure lamp (uviblox® WTP; 2x4 Watt; IBL, Heidelberg, D) was used. During oxidation experiments with hydrogen peroxide the lamp was switched on 10 min prior to the addition of 0.3 g/L H_2O_2. Again the samples were taken time-dependently and catalase (0.25 mL; c = 0.4 kU/mL) was added to terminate the reaction.

4.4.2.4. UV and UV/H_2O_2 oxidation using the flow through system (IUTA e.V.)

The IBL uviblox® WTP 2x4 flow-through unit (IBL, Heidelberg, D) equipped with two 4 kW medium pressure lamps was used for oxidation experiments. For this purpose the water sample was stored in a 1 m^3-IBC-Container and directly pumped into the system passing the UV lamps and dispatched back into the container.

4.4.2.5. UV and UV/H_2O_2 oxidation at the flow through system (Waste water treatment plant)

The above described system was also placed at a waste water treatment plant taking the water directly from the effluent of the plant and releasing it into the discharge system. When H_2O_2 was added to the water an amount of 1 L/h (c = 35 %) was pumped into the system. The flow-through as well as the power of the UV lamps varied (3 – 12 m^3, 1x0.8 – 2x4 kW).

4.5. Extraction methods

4.5.1. Solid Phase Extraction (SPE)

Chemical analyses were performed either using the filtered undiluted non-enriched sample or samples after solid phase extraction. Prior to extraction the samples were filter sterilized using a 1 µm glass-fiber filter (Pall Life Science, Washington, NY, US). Extraction was done using an automated Gilson-System (Valvemate® II; Gilson International B.V., Limburg, D) with Strata X (Phenomenex, Aschaffenburg, D), Strata XL (Phenomenex, Aschaffenburg, D), and ENV+ (Biotage, Uppsala, SE) cartridges. Before extraction, the samples were adjusted to pH 3 and internal standards were added. After extraction the samples were concentrated under nitrogen gas. Pharmaceuticals were eluted with 1 mL water:ACN (50:50) containing 1 % formic acid. Extracts containing Bisphenol A were concentrated under nitrogen gas, eluted using 1 mL methanol, again concentrated and eluted in 100 µL Bis-(trimethylsilyl)-trifluoro-acetamide with 1 % trimethylchlorosilane, mixed and derivatized at 70 °C for 30 min. The extracts of the three organophosphates (TPP, TCEP, TCPP), HHCB and AHTN were stored at -20 °C for 2 h to remove remaining water and afterwards concentrated using a Büchi-Syncore system (Büchi, Flawil, CH) with 60 °C, 40 mbar and 180 rpm (organophosphates) or 50 °C, 60 mbar and 180 rpm (HHCB, AHTN) resulting in a final volume of 1 mL. In addition to the concentration of the organophosphates, the solvent was changed to toluol by adding 10 mL toluol to the sample and concentrating it to 1 mL twice. All samples were then transferred to a vial for measurement. Irgarol 1051 and Terbutryn were transferred to a vial directly after extraction without any further handling.

4.5.2. Liquid-liquid-extraction (LLE)

The three organophosphates as well as the biocides Irgarol 1051 and Terbutryn were extracted using a liquid-liquid-extraction method. Therefore 2 mL of each sample were mixed with 2 mL toluol and 100 µL internal standard was added. After mixing the sample and the toluol for 20 min at 500 rpm the sample was stored at -20 °C for at least two hours. 1 mL of the toluol phase was then aspirated and transferred to a GC-vial for further testing.

Material and Methods

4.6. Analytical Chemistry

Part of the analytical chemistry is the qualitative and quantitative determination of compounds as well as the identification of structures through separation processes. These processes include chromatographic as well as spectroscopic methods. A linkage of both results in a more sensitive analytical tool which allows an improvement in the detection and identification of substances.

Chromatography

During chromatographic separation processes the substance to be analyzed is dissolved in the mobile phase (liquid, gas) and moved over the stationary phase (column, flat-bed). The separation then takes place due to interactions of the compound with the stationary phase resulting in a chromatogram. As the name implies, the gas chromatography is based on a gas used as the mobile phase. After evaporation the sample is then moved across the stationary phase by this gas resulting in a time dependent detection of each compound (retention time). Another type of chromatography is the high performance liquid chromatography (HPLC). The HPLC is a chromatography method using a liquid mobile phase [152].

Mass Spectroscopy

During mass spectroscopy the compounds to be measured are transformed into fast moving, ionizing gases and then detected according to the mass to charge ratio (m/z). This ratio deflects the separation of the ions in an electric or magnetic field [152].

4.6.1. HPLC-MS/MS

A HPLC-MS/MS method was performed to quantify the analytes with the API 3000 (CTC Analytics, Zwingen, CH) and a Q Trap 3200 (AB Sciex, Darmstadt, D) using a Multiple Reaction Monitoring (MRM) mode. Two characteristic mass changes were detected for each substance (Table 24). Chromatographic division was done using a

Material and Methods

Synergi Polar RP 80 HPLC column (150 x 2 mm; 4 µm; Phenomenex, Aschaffenburg, D) on both systems. Three different eluents were used: (1) water with 0.1 % HCOOH and ACN with 0.1 % HCOOH, (2) water with 0.1 % HCOOH and methanol with 0.1 % HCOOH as well as (3) water with 4 mM ammonium acetate (pH 4) and acetonitrile with 0.1 % HCOOH.

Table 24: Mass changes of the analytes during HPLC-MS/MS

Compound	HPLC-MS method	Rentention time [min]	Q1 Mass [Da]	Q3 Quantification [Da]	Q3 Verification [Da]
Ciprofloxacin	Q Trap 3200	2.31	332	314	288
Ethinylestradiol	Q Trap 3200	4.80	297	107	77
Irgarol 1051	API 3000 Pharm. (pos)	2.97	254	198	83
Metoprolol	API 3000 Pharm. (pos)	7.40	268	116	77
Ofloxacin	Q Trap 3200	2.17	362	318	261
Sulfamethoxazole	API 3000 Pharm. (pos)	8.70	254	156	92
Sulfamethoxazole	Q Trap 3200	2.27	254	156	92
Terbutryn	API 3000 Pharm. (pos)	2.71	242	186	68

Pharm. = Pharmaceutical ; pos = positive mode

API 3000:

For the measurements using the API 3000 system the column temperature was set to 35 °C and the pharmaceuticals were quantified in positive and negative mode. Settings for either mode are presented in Table 25.

Material and Methods

Table 25: Measurement settings for pharmaceutical quantification using the API 3000 system

	Positive mode	Negative mode
Flow-through	0.35 mL/min	0.3 mL/min
Nebulizer	12 V	12 V
Curtain gas	10 V	10 V
Ion source temperature	450 °C	450 °C
Ionization voltage	5000 V	-4500 V

Q Trap 3200:

The column temperature of the Q Trap system was set to 40 °C and the samples were transferred to the MS via electron spray ionization (ESI). The settings were as followed: Curtain gas = 15 V, Ion source temperature = 550 °C, Gas 1 = 40 psi, Gas 2 = 80 psi, Ionization voltage = 5500 V, Declustering potential = 50 V, Entrance potential = 6 V, Cell entrance potential = 18 V, Collision energy = 27 eV, Cell exit potential = 4 V. The flow-through was set to 0.4 mL/min for sulfamethoxazole and 0.3 mL/min for the fluoroquinolones.

The analyses were then done with the Analyst 1.5 (AB Sciex; Darmstadt, D) using an external calibration for direct measurements and an internal calibration for previously processed samples. The calibration was weighted 1/x with linear regression, the limit of detection was defined with a signal to noise ratio of 3:1 and the limit of quantification was defined with a signal to noise ratio of 10:1.

In addition to the quantification of known micropollutants a method was developed to screen the samples for unknown micropollutants as well as oxidation by-products. A linear gradient was used for chromatographic division and the detection was carried out with an Information Dependent Acquisition (IDA) experiment. For this purpose, masses between m/z = 50 and 500 were scanned and the ion trap was set to 1000. The following system settings were chosen for the experiment: Collision energy = 10 V, Mass tolerance = 250 mmu. Isotopes with an area of 4.9 Da were excluded and the settings for the gases were the same as in the quantification

Material and Methods

experiments. During each IDA experiment the two masses with the highest intensity were filtered and further fractioned to gain information on the molecular structure of the unknown substances.

4.6.2. GC-MS

Bisphenol A was analyzed with a GC-MS (Trace GC Ultra coupled to a DSQ Mass spectrometer; Thermo-Scientific, Dreieich, D) using a Restek Rxi – 5Sil-MS column (30 m x 0.25 mm, 0.25 µm; Restek, Bellefonte, PA, US). Helium 5.0 (Air Liquide, Düsseldorf, D) was used as carrier gas and the oven temperature was set to 100 °C. The heating rate was 40 °C/min until 260 °C were reached. Then the heating rate was adjusted to 5 °C/min until reaching 310 °C. A DB-5MS column (15 m x 0.25 mm, 0.25 µm; J&W Scientific, Folsom, US) with a linear temperature setting (increasing from 70 °C to 280 °C in 30 min) was used for the analyses of the organophosphates, Triclosan, HHCB and AHTN.

The samples were then vaporized using a Programmed Temperature Vaporizer- (PTV) cooled injection system and transferred to the MS device. The injection volume was set to 1 µL and Helium 5.0 was used as carrier gas (Flow-rate = 1.3 mL/min). Between 100 °C and 320 °C the heating rate was 14.5 °C/min and the temperature of the ion source was 200 °C (Bisphenol A) or 230 °C (Organophosphates, Triclosan, HHCB and AHTN). The Selected Ion Monitoring (SIM) method was used for measurements. Mass transfers for the quantification are depicted in Table 26. The Xcalibur software from Thermo-Scientific (Dreieich, D) was used for the analysis, including a linear regression. The limit of detection was defined with a signal to noise ratio of 3:1 and the limit of quantification was defined with a signal to noise ratio of 10:1.

Material and Methods

Table 26: Mass transfer of the analytes during GC-MS measurements

Compound	1. Mass [Da]	3. Mass [Da]
AHTN	243	258
Bisphenol A	357	372
HHCB	243	258
TCPP	277	279
TBEP	199	299
TPP	325	326
Triclosan	288	290

4.7. Peroxide testing

The testing for remaining peroxides after UV/H_2O_2 treatment was done using peroxide tests strips (Quantofix®) which allow a semiquantitative analysis (Figure 27) based on a color scale. With these test strips peroxides were detected at concentrations between 0.5 and 25 mg/L.

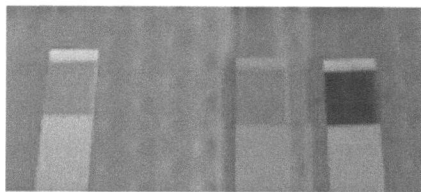

Figure 27: Peroxide Test-strips (Quantofix®). Left: negative for peroxides; right: peroxides detected

Material and Methods

4.8. Waste water treatment plant effluent used for testing

The here used effluent was collected at a waste water treatment plant which is able to clean up to 8500 L/sec and which is designed for a capacity of 1.22 million population equivalents serving five cities with a population between ca. 75,000 and 574,635 persons. It was build as a conventional treatment plant consisting of a mechanical treatment (rack, sand and grease removal through sedimentation), an activated sludge basin, and primary as well as secondary treatment steps. Up to now, advanced or oxidative treatment step are not applied before the water is released.

The DOC of the here used WWTP effluent ranged from 6 to 8 mg/L over the course of the experiments and the pH was between pH 7 and pH 8.5. Chemical analyses revealed that the WWTP effluent still contained micropollutants, however at concentrations below 1 µg/L.

4.9. Sample preparation before toxicological testing

The water samples had to be sterilized by filtration before they were added to the cells in order to prevent bacterial contamination. For this reason sterile filters with a pore size of 0.2 µm were used.

5. Results

5.1. Adaptation of toxicological methods

5.1.1. MTT Test, PAN I and Alkaline Comet Assay

The MTT test, the PAN I test and the alkaline comet assay are methods routinely used for the detection of toxic effects of chemicals and other substances, e.g. nanoparticles or pharmaceuticals. However they are not suitable for the testing of water samples since the addition of water to the cells might lead to an increase in the osmotic pressure and subsequent cell death.

In order to apply the mentioned tests, the MTT test and the Alkaline Comet Assay had to be adapted to the matrix since they are not applicable for the testing of water samples and to prevent false positive and false negative results.

The first step therefore was the determination of the highest water concentration that can be used for an exposure without causing toxic effects leading to false positive results. Positive controls were monomethylarsonous acid (MMA III; 50 µM) for the MTT test and 0.1 mg/mL N-ethyl-N-nitrosourea (ENU) for the Alkaline Comet Assay. The testing of a dilution series with different amounts of water added to the exposure medium showed that a final concentration of 10 % water does neither lead to cytotoxic nor genotoxic effects in CHO or T47D cells (Table 27) whereas the smaller dilution factor (1:2) results in strong toxic effects in the MTT test and highly significant DNA damages in the Alkaline Comet Assay. Based on these results a dilution of 1:10 of the water samples into the exposure medium was then further used during the course of all experiments.

Results

Table 27: Toxic effects of different amounts of water in exposure medium

Sample [dilution]	MTT Test [degree of cytotoxicity]		Alkaline Comet Assay [statistical significance of DNA damage]
	CHO cells	T47D cells	CHO cells
Neg. control	0	0	not significant
Pos. control	3*	3*	highly significant[#]
1:2	3	3	n.t.
1:10	0	0	not significant
1:20	0	0	not significant

n.t. = not tested; * = 50 µM MMA III; # 0.1 mg/mL ENU; not significant = $p > 0.05$; highly significant = $p < 0.001$

5.1.2. ER Calux

The ER Calux is a test which can be used for the testing of water samples however the directions provided by BDS include an extraction step followed by an elution in DMSO. To avoid this step in order to test the water sample non-enriched for a realistic statement the method was optimized. Therefore it was first determined whether a white or a black 96-well plate should be used for the luminescence measurement. A concentration series of a Luciferase standard was used after adding the GlowMix to the solution. The results show that the relative light units (RLU) of the Luciferase standard in the white plate resulted in a coefficient of determination (R^2) of 0.94 whereas the measurement of the black plate resulted in a R^2 of 0.71 (Figure 28). Based on this data the white plate was then used for the RLU measurements.

Results

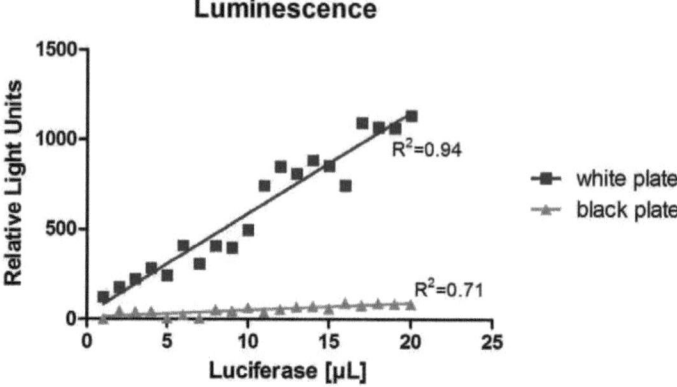

Figure 28: Luminescence measurement of a Luciferase standard using a white and a black 96-well plate.

5.1.3. Ames Test

The Ames MPF™ 98/100 Aqua in contrast has been developed for the testing of water samples. Thus there was no need for adaptation or optimization and the test was performed according to the manual provided by Xenometrix (see section 2.6.5 Ames Test).

5.1.4. H_2O_2

Due to the use of hydrogen peroxide (H_2O_2) during UV treatment a proportion of peroxides might still remain in the sample because of an incomplete reaction. When the cells are then exposed to water samples still containing these remaining peroxides the sample might be toxic leading to false positive results. To overcome these effects catalase was added to the oxidized samples to degrade remaining peroxides before exposure.

Results

Catalase is an enzyme which reduces the toxic H_2O_2 to oxygen and water [153]:

$$2\ H_2O_2 + \text{Catalase} \rightarrow O_2 + 2\ H_2O$$

Preliminary tests showed that UV/H_2O_2 treated samples had both cytotoxic (Figure 29a) and genotoxic (Figure 29b) effects. In addition the test for peroxides was positive.

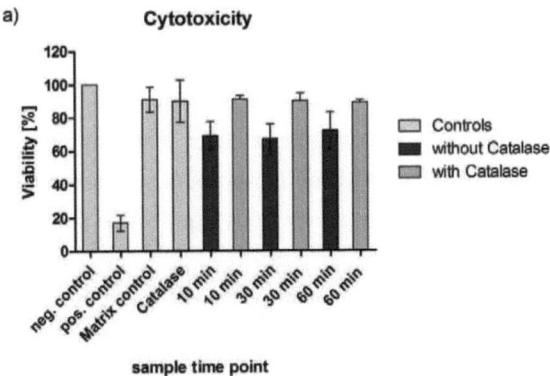

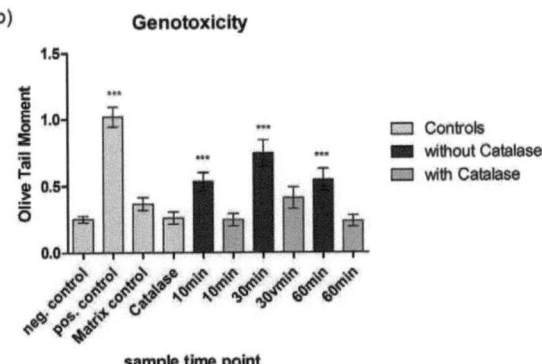

Figure 29: a) Cytotoxic effects of UV/H_2O_2 treated samples. b) DNA damaging effects of UV/H_2O_2 treated samples. The addition of catalase resulted in the elimination of toxic effects.

Results

Samples treated with UV and H_2O_2 resulted in a loss of viability and were classified as moderately (10 min and 30 min) or weakly (60 min) cytotoxic. The statistical analysis of the Olive Tail Moments (OTM) showed highly significant increases in the OTM of the treated sample compared to the negative control. However the addition of 0.25 mL catalase (0.4 KU/mL) to the toxic samples resulted in a complete elimination of peroxides and thus toxic effects (Figure 29). Catalase itself did not result in cytotoxic or genotoxic effects and neither did the matrix control. Therefore catalase was then added to all H_2O_2 treated samples and they were then in addition checked for remaining peroxides to rule out false positive results.

5.1.5. Contaminations

In regard to the testing of waste water treatment plant effluents bacterial contamination has to be avoided. Filtration has been proven as a sufficient method. As shown in Figure 30 the samples were spread on blood agar plates before and after sterile filtration. The sample plated without filtration shows a heavy microbial contamination (Figure 30a) whereas the filtered sample does not result in microbial growth (Figure 30b). Further on all waste water samples were therefore filtered before exposure.

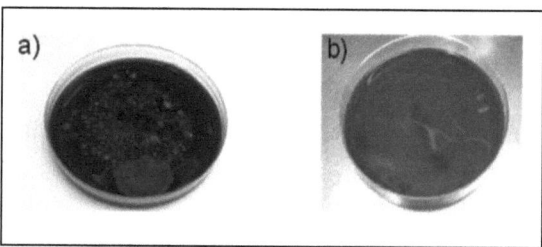

Figure 30: Blood agar plates with bacterial growth after plating an unfiltered WWTP effluent sample (a) and a plate without colonies after plating a filtered sample (b) incubated for 24 h at 37 °C.

Results

5.2. Matrix controls

For all tests either HPLC-water or waste water treatment plant effluent was used to dissolve the substances for oxidative treatment and further chemical and toxicological analysis. Therefore these waters were first tested for cytotoxicity and genotoxicity before and after treatment to prevent matrix effects through the oxidation methods or in the case of the WWTP effluent through substances present in the water.

5.2.1. HPLC-water

The HPLC-water used to dissolve the substances was first tested for toxic effects to rule out possible false positive or negative effects due to the water matrix.

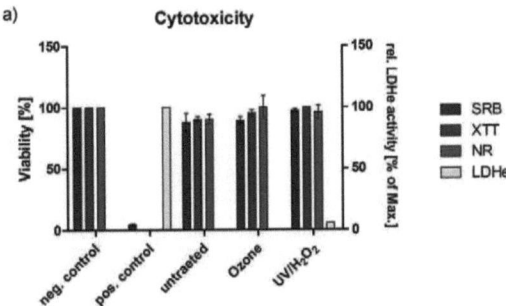

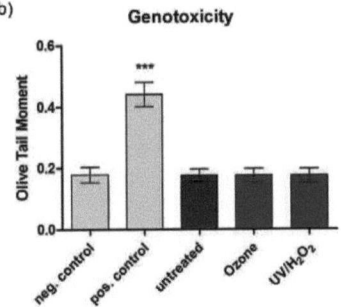

Figure 31: No cytotoxicity (a) and genotoxicity (b) of HPLC-water before and after ozonation (2 mg O_3/L) or UV/H_2O_2 treatment.

Results

Therefore the HPLC-water was tested before and after ozonation (5 mg/L O_3) or UV/H_2O_2 (Hg-LP, 15 W; 1 g/L H_2O_2) treatment. Neither test (PAN I Multitox test or Alkaline Comet Assay) revealed toxic effects when compared to the negative control (Figure 31).

To rule out false positive effects through methodological errors during extraction, extracts of HPLC-water (C18 and Strata X) were also tested for toxic effects (Figure 32). The results of the Alkaline Comet Assay show that neither the extracted nor the original HPLC-water samples had a statistically significant influence on the Olive Tail Moment compared to the negative control. There was also no difference of the extracted samples compared to the original HPLC-water sample. In addition the ozonation also did not have an influence on the genotoxicity of the samples.

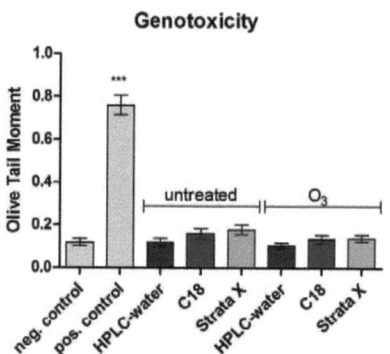

Figure 32: No genotoxicity of HPLC water extracts (C18 and Strata X) before and after 60 mintes of ozonation

Results

5.2.2. Waste water treatment plant effluent

In addition to the above presented results for HPLC-water, waste water treatment plant effluent was also tested for cytotoxicity and genotoxicity before and after oxidative treatment without the addition of the substances. The results demonstrated that neither the untreated, nor the effluent after ozonation or UV/H_2O_2 treatment resulted in a loss in viability (Figure 33a) or an increase in DNA damage (Figure 33b).

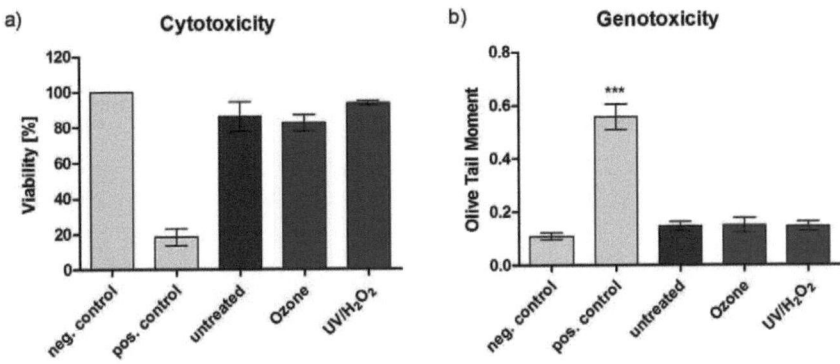

Figure 33: Waste water treatment plant effluent before and after oxidative treatment. No cytotoxic (a) or genotoxic (b) effects.

Extracts of waste water treatment plant effluent using C18 or Strata X columns were then also tested for their potential to induce DNA damaging effects before and after ozonation, to rule out false positive effects due to the extraction procedure. As can be seen in Figure 34 the Olive Tail Moments of both extracts (C18 and Strata X) are comparable to both, the original water sample without extraction as well as the negative control. There is also no statistically significant difference in the Olive Tail Moment between the untreated samples and the samples after 60 minutes of ozonation.

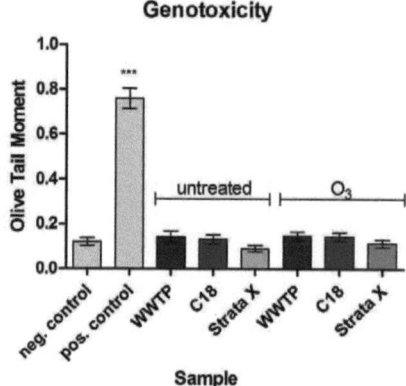

Figure 34: No genotoxicity of WWTP extracts (C18 and Strata X) of WWTP effluent before and after ozonation

5.3. ß-Blocker

Atenolol and Metoprolol both belong to the group of $ß_1$-adrenoreceptor antagonists inhibiting the function of hormones like adrenaline and noradrenaline. Both hormones belong to the group of catecholamines, synthesized in cells of the adrenal cortex and are transported to the heart as a response to stress affecting heart cells by binding to the adrenergic receptors which subsequently will lead to an increase in the heart rate and blood pressure as well as other metabolic functions [116]. Thus ß-blocker are used to treat a variety of diseases compromising the cardiac system [154]. Metoprolol is still the most frequently prescribed ß-blocker and Atenolol the third most one [34]. Their widespread use for many years (e.g. 204,088 kg in 2009 [12]) and an increase of 6.4 % in the amount prescribed in 2010 compared to 2009 [34] has led to their high abundance in the water system.

5.3.1. Atenolol

2 mg/L Atenolol were dissolved in HPLC-water were then tested for toxicity before and after oxidative treatment leading to a concentration of 0.2 mg/L Atenolol during exposure. The results show that neither before nor after ozonation with 2 mg O_3/L cytotoxic (Figure 35a) or genotoxic (Figure 35b) effects occur since the viability of the CHO cells does not decrease below 80 % in the XTT, SRB and NR test. In addition the relative LDHe activity after an exposure to the water samples is comparable to the negative control. The same is true for the genotoxicity testing. The Olive Tail Moment of the exposed CHO cells is in the range of the negative control and no statistically significant changes were detected. Thus the results prove that Atenolol itself is not cyto- or genotoxic at the tested concentration. The degradation analyses of Atenolol after ozonation revealed no formation of oxidation by-products since it was degraded immediately after adding 2 mg/L of ozone to the water sample [151].

Results

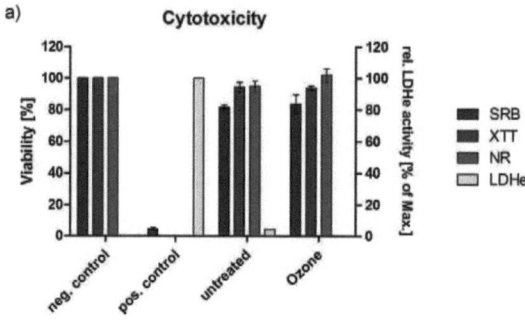

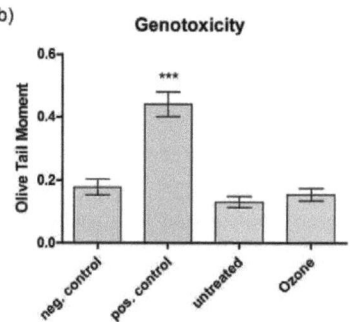

Figure 35: No cytotoxicity (a) and genotoxicity (b) of 0.2 mg/L Atenolol in HPLC-water before and after ozonation (2 mg O_3/L).

5.3.2. Metoprolol

For the ozonation experiments using HPLC-water, 14 mg/L of Metoprolol were added to HPLC-water and subjected to 5 mg/L of ozone for 60 minutes. No cytotoxicity (Figure 36a) or genotoxicity (Figure 36b) of 1.4 mg/L Metoprolol were detected and no differences were seen before (untreated) or after (O_3) ozonation. The same results were detected for HPLC-water containing 0.1 mg/L Metoprolol before and after UV/H_2O_2 treatment, where also no toxic effects on the cell viability were detected in the MTT test (Figure 36a) or damages to the DNA using the Alkaline Comet Assay (Figure 36b).

Results

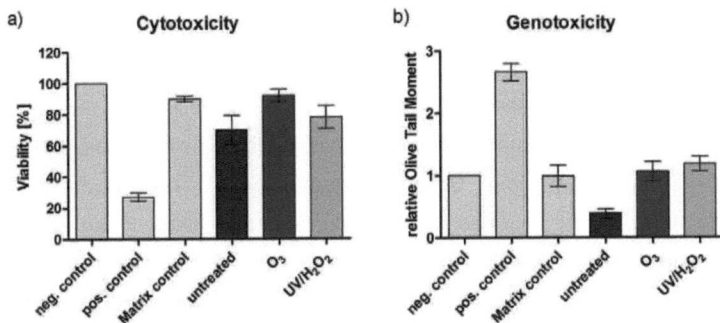

Figure 36: a) No cytotoxicity or genotoxicity (b) of HPLC-water containing 1.4 mg/L Metoprolol before and after ozonation or 0.1 mg/L Metoprolol before and after UV/H$_2$O$_2$ treatment for 60 minutes.

The same concentrations of Metoprolol were then also tested for toxic effects when added to waste water treatment plant effluent and then oxidatively treated for 60 minutes. No decrease in viability was detected before and after ozonation or UV/H$_2$O$_2$ treatment (Figure 37a). The Alkaline Comet Assay also did not show any influence of the tested samples on the Olive Tail Moment as the measure for DNA damage (Figure 37b).

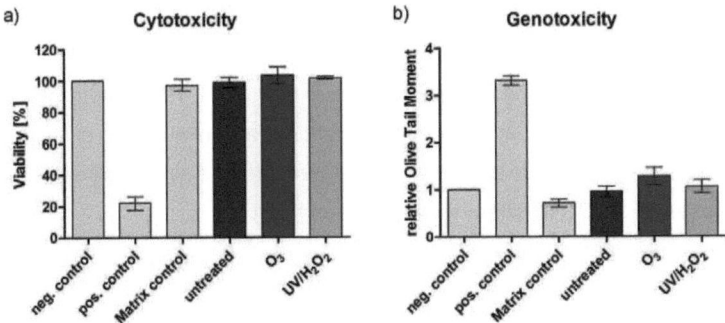

Figure 37: a) No cytotoxicity or genotoxicity (b) of WWTP effluent with 1.4 mg/L Metoprolol before and after ozonation or of 0.1 mg/L Metoprolol after UV/H$_2$O$_2$ treatment for 60 minutes.

5.4. Estrogenic substances

Estrogens like estradiol or estrone are hormones, promoting the development of the female secondary sex characteristics. Besides this they play a role e.g. during osteogenesis or the lipid metabolism [116]. Their mode of action is based on the diffusion through the cell membrane followed by binding to an estrogen receptor only present in specific tissues (e.g. breast, uterus, ovaries). The now activated estrogen:receptor complex migrates into the nucleus and binds to an estrogen responsive element of the DNA initiating DNA transcription and the production of proteins. Naturally, estrogens are synthesized in vertebrates and also some invertebrates through the transformation of male sexual hormones. Synthetic estrogens, also known as xenoestrogens, are present as chemicals, plasticizers (e.g. Bisphenol A), pharmaceuticals (e.g. 17α-ethinylestradiol) and cosmetics [155]. Although the use of estrogens and xenoestrogens in medicine is widespread the amount of prescribed pharmaceuticals containing estradiol slightly decreased (0.9 %) in 2010 compared to the amount prescribed in 2009 [34].

Synthetic estrogens are the most important compounds described as endocrine disrupting chemicals (EDC). The word endocrine defines the process of the formation of a hormone in an organ and its transport to the target location via the blood [116]. EDC is the term for exogenous substances which mimic the action of natural hormones thus affecting the hormonal system by interfering with natural signaling pathways resulting in a variety of possible health effects [156, 157].

The most prevalent effects of EDC in the water system have been detected in fish. In order to develop the female type estrogens are essential [116]. Thus if estrogens are present in the environment at high concentrations a feminization of this fish population might take place. Despite this, *in vitro* test sytems have also been able to show the presence of EDC in different water bodies [158].

Results

5.4.1. Ethinylestradiol

Ethinylestradiol (EE2) is a synthetic derivative of the natural estrogen estradiol and is mainly used as a contraceptive [34]. Since it is one of the most used pharmaceuticals, the consumption of pharmaceuticals containing Ethinylestradiol was almost 633 kg in 2009 [12]. Compared to estradiol the bioavailability of ethinylestradiol is increased due to the substitution of an ethyl-group at the C17 position resulting in a decreased first-pass-effect in the liver and therefore lower biodegradability and a higher excretion rate of the unmetabolized molecule [34].

First 15 mg/L Ethinylestradiol were dissolved in HPLC water and subjected to UV radiation with H_2O_2 and treated for up to 60 minutes. In addition the same concentration of EE2 was also treated with 1, 5 or 10 mg/L of ozone for 10 or 60 minutes. Both sets of samples were then first tested for cytotoxic effects against T47D cells with a final concentration of 1.5 mg/L EE2 during exposure. As can be seen in Figure 38 neither the untreated EE2 sample nor the UV/H_2O_2 treatment (Hg-LP, 15 W; c(H_2O_2) = 0.3 g/L) for 10 or 60 minutes did result in cytotoxic effects. In addition the ozonation using three different concentrations of ozone (1, 5, and 10 mg/L) did also not lead to a decrease in the number of viable T47D cells below 80 %, thus there were also no cytotoxic effects detected for ozonated samples.

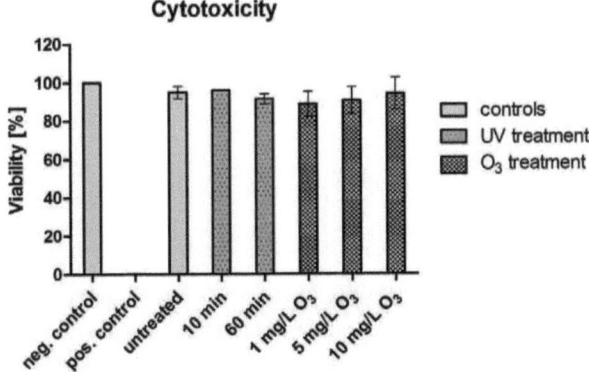

Figure 38: No cytotoxicity of 1.5 mg/L Ethinylestradiol before and after UV/H_2O_2 treatment or ozonation (1-10 mg/L O_3).

Results

Since Ethinylestradiol is a synthetic estrogen the samples were also tested for the induction of estrogenic effects before and after oxidative treatment using the ER Calux. Displayed in Table 28 are the results of the testing before and after ozonation or UV/H_2O_2 treatment of HPLC-water containing 1.5 mg/L Ethinylestradiol. The EEQ values clearly show that there is a decrease in estrogenicity after both treatment methods compared to the untreated sample. An increase in treatment duration (UV/H_2O_2) and ozone concentration leads to a decrease in estrogenic activity. In fact the UV/H_2O_2 treatment resulted in a lower estrogenic activity after 60 min than the highest ozone concentration (10 mg/L) after 60 minutes. However both methods did not result in a complete elimination of the estrogenic activity from the water sample and only slight differences in EEQ values were seen after treatment.

Table 28: Estrogenicity of 1.5 mg/L Ethinylestradiol dissolved in HPLC-water before and after UV treatment and ozonation.

Sample	Estrogenicity (ER Calux EEQ [pM/Well])	Estrogenicity (ER Calux EEQ [ng/L])
untreated	1.8	4.90
10 min UV*	1.4	3.81
60 min UV*	0.9	2.45
1 mg/L O_3 [#]	1.6	4.36
5 mg/L O_3 [#]	1.3	3.54
10 mg/L O_3 [#]	1.1	3.00

* Hg-LP, 15 W; c(H_2O_2) = 0.3 g/L; [#] duration = 60 min

5.4.2. Bisphenol A

According to the European Food Safety Authority, Bisphenol A is a chemical substance mainly used as a basic material for the synthesis of polycarbonates and it is also used as an antioxidant in plasticizers. These polycarbonates are then used in everyday products e.g. storage boxes and plastic bottles. Due to its widespread use (global production = 3.8 million tonnes in 2006 [159]), it has been detected in surface

Results

waters and waste water treatment plant effluents and thus might pose a threat to the environment because of its endocrine disrupting mode of action.

The toxicity of Bisphenol A was tested before and after oxidative treatment of waste water treatment plant effluent or HPLC-water. The results of HPLC-water containing 1.4 mg/L Bisphenol A show no cytotoxic effects after ozonation (O_3 = 5 mg/L), whereas 0.1 mg/L Bisphenol A resulted in a decrease in viability indicating strong cytotoxic effects and the formation of toxic oxidation by-products (Figure 39a) after 60 minutes of UV oxidation (Hg-LP lamp: 15 W) in combination with H_2O_2 (c = 1 g/L). In addition no increase in Olive Tail Moment was detected for the ozone treated sample, while the cytotoxic sample was not tested for genotoxic effects (Figure 39b). Thus no genotoxicity could be detected for the used concentrations.

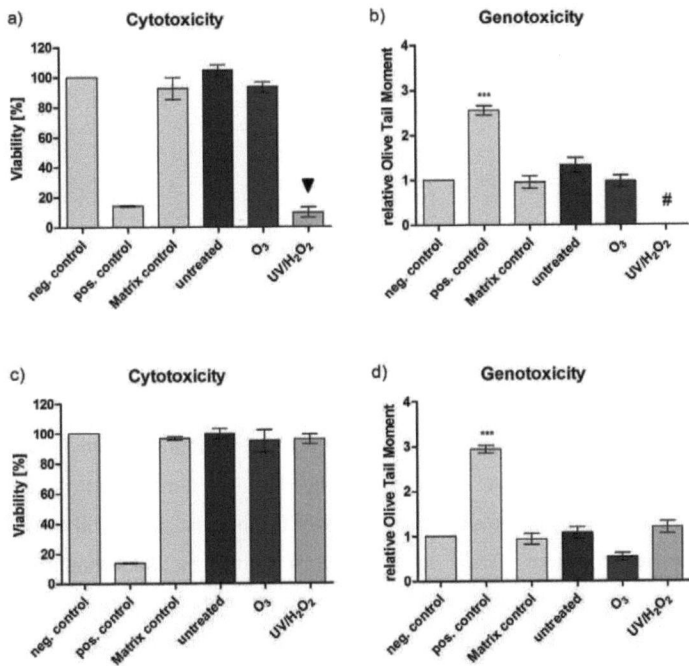

Figure 39: a) No cytotoxic effects of HPLC-water containing 1.4 mg/L Bisphenol A after ozonation, but cytotoxic effects after UV/H_2O_2 treatment. b) No genotoxic effects of tested samples. No cytotoxic (c) or genotoxic (d) effects of WWTP effluent containing Bisphenol A before and after ozonation or UV/H_2O_2 treatment. ▼ cytotoxic effects ;# = not tested

Results

In addition to the cytotoxicity and gentoxicity testing 0.1 mg/L Bisphenol A in HPLC-water were also tested for the induction of estrogenic effects before and after UV/H_2O_2 treatment. EEQ values were only detected in the untreated sample, whereas 10 min or 60 min resulted in a complete loss in estrogenicity (Table 29).

Table 29: EEQ values [pM/100 µL] of 0.1 mg/l Bisphenol A in HPLC-water

UV/H_2O_2 treatment (Hg-LP, 15 W; 1 g/L H_2O_2)	ER Calux EEQ [pM/100 µL]	ER Calux EEQ [ng/L]
Untreated	2.7	5.72
10 min	-*	-*
60 min	-*	-*

* = no estrogenicity detected

5.5. Antibiotics

Antibiotics are natural substances synthesized mostly by fungi as metabolites and are selectively effective against other microorganisms, e.g. bacteria [160, 161]. The first antibiotic (Penicillin) was isolated from a broth culture of the mold *Penecillium notatum* (now *Penicillium chrysogenum*) and described by Alexander Fleming in 1929 [162]. Since that time many more antibiotic substances have been isolated and are commercially synthesized for the use in medicine. Antibiotics are effective by e.g. inhibiting the cell wall synthesis, interfering with the membrane permeability, inhibiting the protein synthesis, metabolic functions or the nucleic acid synthesis and replication of microorganisms. They can therefore either be bacteriostatic by inhibiting proliferation or bactericidal by killing these organisms. Due to their mode of action antibiotics are classified into different groups (ß-lactams, sulfonamids, quinolones, cephalosporines etc.) [160, 161]. For the years 2008 and 2009 a total amount of approximately 374 million DDD of antibiotics and 41 million prescriptions for ambulant therapies excluding hospital cases were listed. This numbers make antibiotics one of the top three pharmaceutical groups prescribed annually [163].

Results

In this project the antibiotics Sulfamethoxazole, Ciprofloxacin, Ofloxacin, and Triclosan were investigated for their toxicological effects before and after oxidative treatment.

5.5.1. Sulfamethoxazole

Sulfamethoxazole belongs to the group of sulfonamide antibiotics and is used for the treatment of e.g. urinary tract infections and pneumonia [160]. It is always applied in combination with Trimethoprim, another bacteriostatic agent [164]. Sulfonamide antibiotics work by inhibiting the 4-Aminobenzoic acid, an intermediate during the synthesis of folate, which in the end leads to interruptions of DNA and RNA synthesis and thus a disturbed synthesis of amino acids, proteins and nucleotides. It can be used in human medicine since humans do not synthesize folic acid themselves and therefore will not be affected by this group of antibiotics [160]. The sum of Sulfamethoxazole containing antibiotics added up to 15.8 million DDD in 2011 which is a decline of 3.5 % compared to 2009. In fact sulfonamide antibiotics are the group with the lowest number of prescriptions in 2011 compared to other groups of antibiotics, e.g. fluoroquinolones or makrolides [34].

As can be seen in Figure 40 the viability of CHO cells exposed to HPLC-water containing 0.1 mg/L Sulfamethoxazole before and after UV/H_2O_2 treatment (Hg-LP lamp, 15 W; 1 g/L H_2O_2) did not result in a decrease of the viability below 80 %, and therefore no cytotoxicity was detected. The Olive Tail Moment of the CHO cells exposed to the same samples did also not increase significantly in comparison to the negative control. Thus there are neither cytotoxic nor genotoxic effects of Sulfamethoxazole itself and its oxidation by-products in HPLC-water.

Results

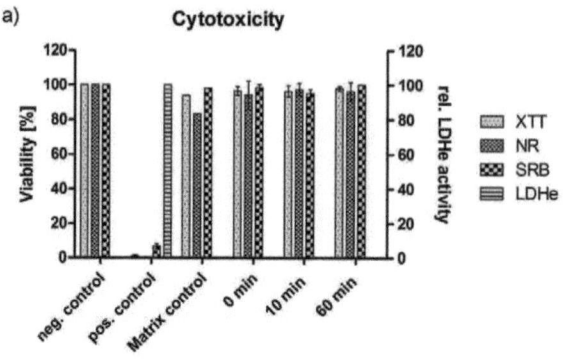

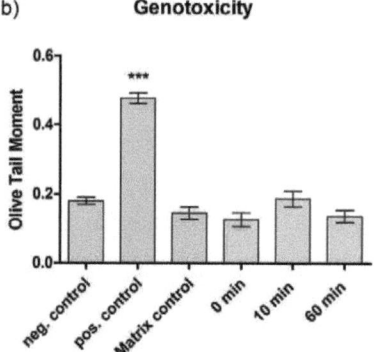

Figure 40: Sulfamethoxazole (0.1 mg/L) in HPLC-water before and after UV/H$_2$O$_2$ treatment. No cytotoxicity (a) or genotoxicity (b) detected.

The ozonation (5 mg/L O$_3$) of HPLC-water containing 14 mg/L Sulfamethoxazol did also not lead to the detection of cytotoxic (Figure 41a) or genotoxic (Figure 41b) effects in CHO cells after an exposure to 1.4 mg/L Sulfamethoxazole. There was no significant difference between the untreated (0 in) and treated (60 min) samples.

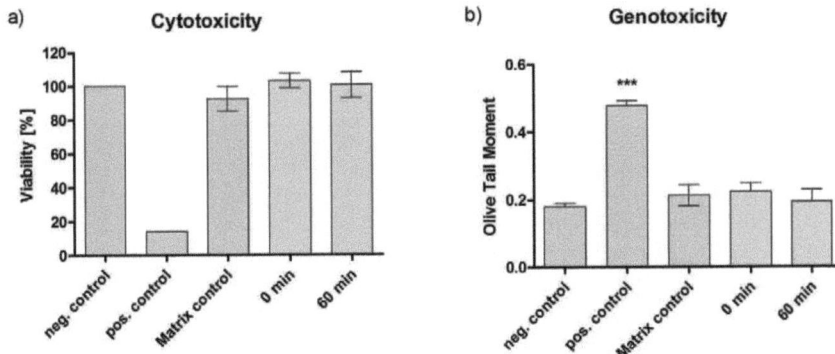

Figure 41: Sulfamethoxazole (1.4 mg/L) in HPLC-water before and after ozonation. No cytotoxicity (a) or genotoxicity (b) detected.

Sulfamethoxazole was then dissolved in WWTP effluent (14 mg/L) and also tested for cytotoxic and genotoxic effects before and after ozonation using 5 mg/L O_3. The results of the PAN I cytotoxicity test revealed no effects of 1.4 mg/L Sulfamethoxazole, neither before nor after ozonation (Figure 42a). The same samples did also not result in a significant increase in DNA damage (Figure 42b) thus no toxicity was detected with the used test systems.

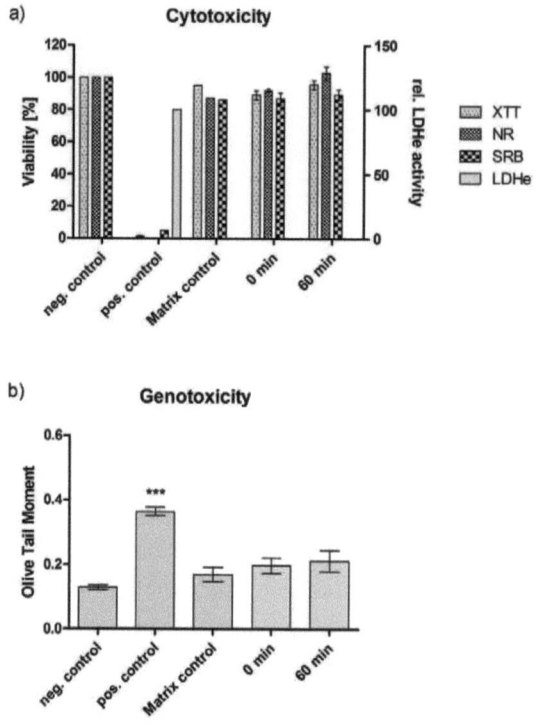

Figure 42: No cytotoxicity (a) and genotoxicity (b) of 1.4 mg/L Sulfamethoxazole in WWTP effluent before (0 min) and after (60 min) ozonation (5 mg/L O_3).

Finally, 0.1 mg/L Sulfamethoxazole were tested for cytotoxicity and genotoxicity in WWTP effluent before and after UV/H_2O_2 treatment (Hg-LP, 15 W; 1 g/L H_2O_2). The results clearly show that there is no significant decrease in cell viability in all tested samples compared to the negative control (Figure 43a). In addition the same samples of Sulfamethoxazole were tested using the Alkaline Comet Assay and no genotoxicity was detected for 0.1 mg/L Sulfamethoxazole itself, (0 min) and the formed oxidation by-products (60 min).

Results

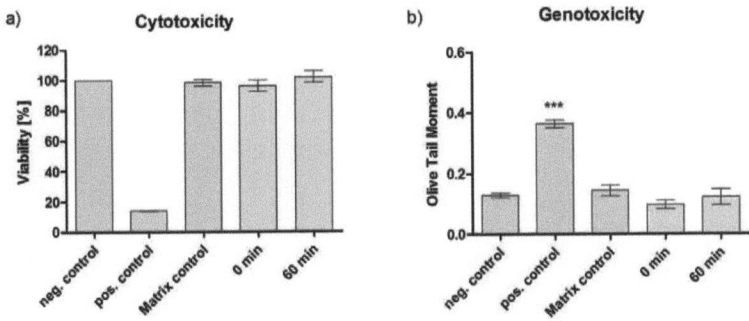

Figure 43: No cytotoxicity (a) and genotoxicity (b) of 0.1 mg/L Sulfamethoxazole in WWTP effluent before and after UV/H_2O_2 treatment.

5.5.2. Ciprofloxacin

Ciprofloxacin is a second generation quinolone antibiotic. This group of antibiotics contains a fluorine atom thus called fluoroquinolone antibiotics. Fluoroquinolones inhibit the activity of the bacterial enzyme gyrase which is responsible for the supercoiling and packaging of bacterial DNA. Thus fluoroquinolone antibiotics are mostly bactericidal. The use of Ciprofloxacin is mainly during the treatment of urinary tract infections and influenza infections [160, 161]. Between the years 2002 and 2009 the amount of consumed Ciprofloxacin was almost doubled (17183.1 kg in 2002 and 32979.5 kg in 2009) [12] which is also displayed in 17.8 DDD in 2009 with a plus of 9.5 % compared to 2008 [34].

17 mg/L Ciprofloxacin in HPLC-water were treated with UV-light (Hg-LP, 15 W) and H_2O_2 (1 g/L) for up to 60 min. While there was no cytotoxicity of the untreated sample (1.7 mg/L Ciprofloxacin), the UV/H_2O_2 treatment resulted in moderate cytotoxic effects reducing the viability to 69.35 % (Figure 44a). Except the one cytotoxic sample (60 min UV/H_2O_2 treatment) all others were also tested for genotoxicity with an additional sample after 30 min of oxidative treatment. However none of the samples resulted in genotoxic effects (Figure 44b). For a completion of the toxicity testing with Ciprofloxacin in HPLC-water, samples containing 14 mg/L

Ciprofloxacin were also treated with ozone (5 mg/L) and tested for their potential cytotoxicity and their DNA damaging potential.

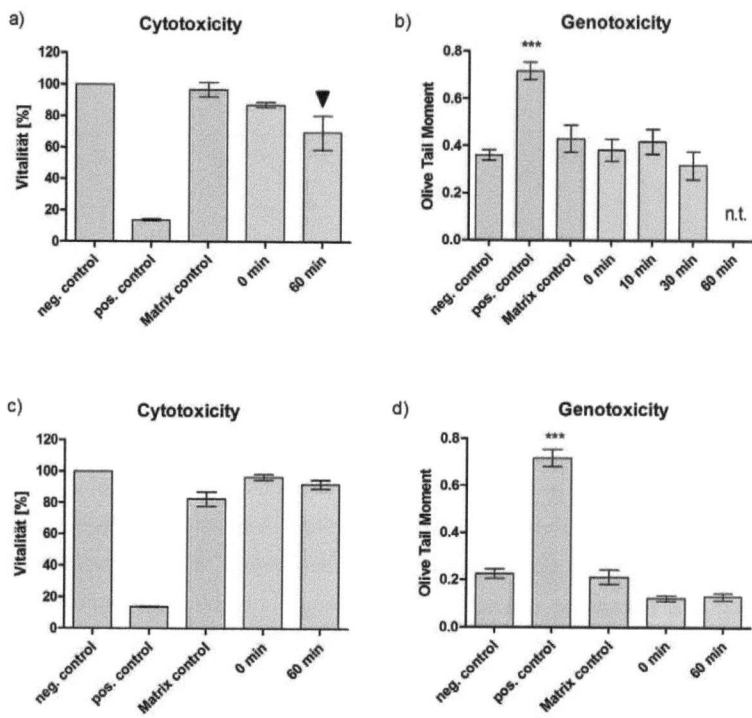

Figure 44: a) Cytotoxic effects of 1.7 mg/L Ciprofloxacin in HPLC-water after 60 min UV/H_2O_2 treatment. b) No genotoxic effects before and after a maximum of 30 min UV/H_2O_2 oxidation. c) No cytotoxic or (d) genotoxic effects of 1.4 mg/L of Ciprofloxacin in HPLC-water before and after ozonation. ▼ Cytotoxic; n.t. = not tested

The results displayed in Figure 44c indicate that neither before nor after the ozonation cytotoxic effects of 1.4 mg/L Ciprofloxacin occur. The same samples did also not induce an increase in DNA damage (Figure 44d).

WWTP effluent spiked with 14 mg/L Ciprofloxacin was subjected to ozonation (5 mg/L O_3) and tested for toxicity after 60 min of treatment. No effects on the viability of the CHO cells was seen (Figure 45a) using the PAN I cytotoxicity test.

Results

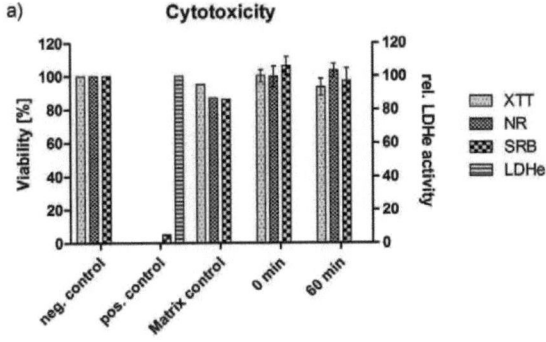

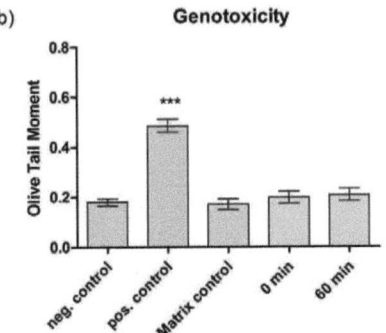

Figure 45: WWTP effluent containing 1.4 mg/L Ciprofloxacin show no cytotoxic (a) and genotoxic (b) effects after ozonation (5 mg/L).

There were also no genotoxic effects of 1.4 mg/L Ciprofloxacin detectable before and after ozonation (Figure 45b). In addition a WWTP sample containing 1 mg/L Ciprofloxacin was also tested for cytotoxicity and genotoxicity (1 mg/L Ciprofloxacin) before and after UV/H_2O_2 oxidation but neither a decrease in cell viability (Figure 46a) nor an increase in the OTM (Figure 46b) was detected, thus no toxic oxidation by-products have been formed.

a)
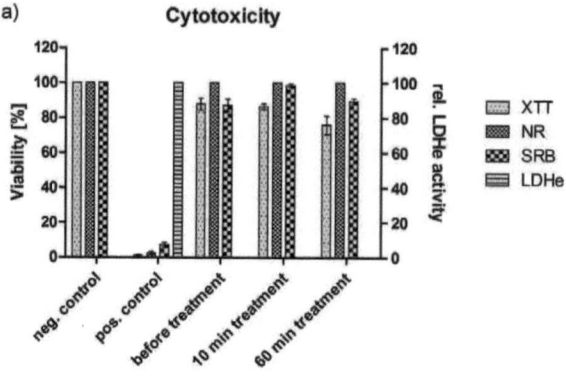

b)
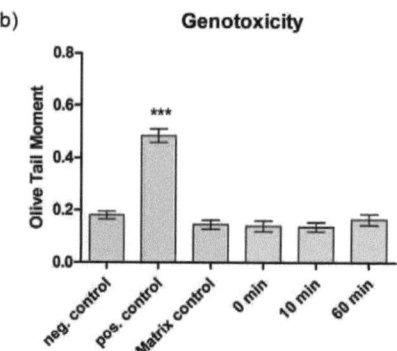

Figure 46: a) No cytotoxic and b) genotoxic effects of untreated and UV/H_2O_2 (Hg-LP, 15 W; 1 g/L H_2O_2) treated WWTP effluent with 0.1 mg/L Ciprofloxacin.

Ciprofloxacin is also thought of as a mutagenic substance since a few studies have been published showing its mutagenic activity [165, 166]. However the results concerning the mutagenicity of Ciprofloxacin are differing since others showed no effects [167, 168]. Therefore the Ames test was also performed with the above presented samples. The results are displayed in Table 30. 1.4 mg/L Ciprofloxacin in HPLC-water did not result in a reversed mutation in the Ames test using the Salmonella typhimurium strains TA98 and TA100. No significant differences were

Results

observed between the untreated (0 min) and the UV/H$_2$O$_2$ treated samples (10, 30 and 120 min). There was also no difference in the number of positive wells detectable comparing the strains tested with and without the addition of the S9 liver homogenate. Thus the metabolism has no influence on the mutagenicity of Ciprofloxacin at the here tested concentrations.

Table 30: **Ames test results of 1.4 mg/L Ciprofloxacin in HPLC-water before and after UV/H$_2$O$_2$ treatment showing the average number of positive wells.**

	TA98 −S9	TA98 +S9	TA100 −S9	TA100 +S9
Negative control	1	1	2,33	1,83
0 min	0	0	0	0
10 min	0	0	0	0
30 min	1,33	1	0	0,33
120 min	0,67	0,67	2,67	0,67
Positive control	29,33	48	29,67	18,67

Results

5.5.3. Ofloxacin

Like Ciprofloxacin, Ofloxacin belongs to the second generation of fluoroquinolone antibiotics and is used to treat infections of the pulmonary tract, the gastrointestinal tract, and the urinary tract system [169]. According to its safety data sheet only little information about toxic effects is available. A total amount of 2 Million DDD (defined daily doses) of Ofloxacin containing antibiotics were prescribed in 2010 [34].

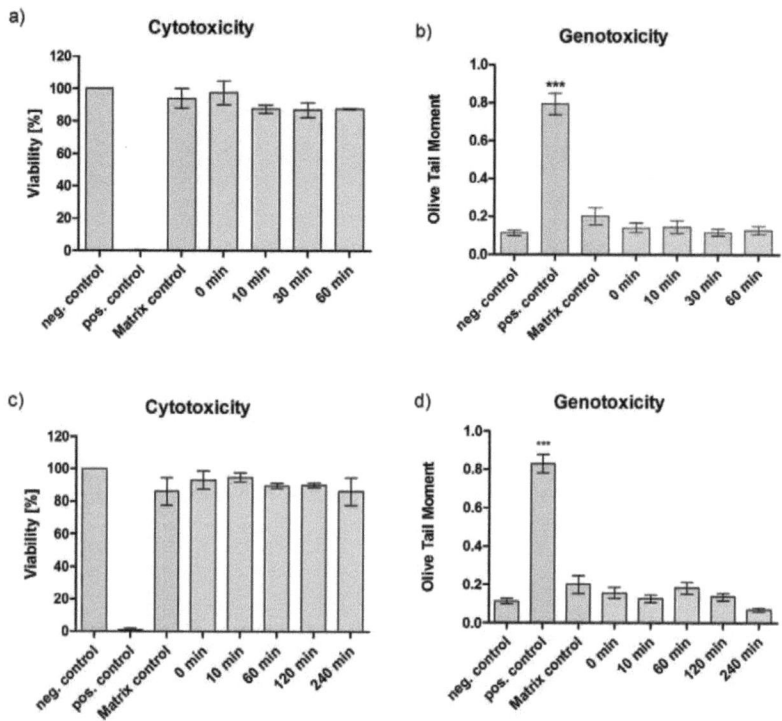

Figure 47: Cytotoxicity (a) and genotoxicity (b) testing of HPLC-water containing 18 mg/L Ofloxacin before and after ozonation (10 mg/L O_3). No cytotoxicity (c) or genotoxicity (d) of 18 mg/L Ofloxacin in HPLC water before and after UV/H_2O_2 treatment.

Results

In this study cytotoxicity and genotoxicity tests have been performed with Ofloxacin in WWTP effluent as well as in HPLC-water. The final Ofloxacin concentration during exposure was 18 mg/L. The exposure of CHO cells to HPLC-water with Ofloxacin before and after ozonation did not result in cytotoxic (Figure 47a) or genotoxic (Figure 47b) effects when treated for up to 60 min. There were also no cytotoxic (Figure 47c) or genotoxic (Figure 47d) effects detectable after UV/H_2O_2 treatment for up to 4 h.

CHO cells were also exposed to WWTP effluent containing 18 mg/L Ofloxacin before and after ozonation and UV/H_2O_2 treatment (both for 60 min). The results displayed in Figure 48 show that there neither is a loss in viability (Figure 48a), nor is there a statistically significant increase in the amount of DNA damage (Figure 48b). Thus no toxic by-products have been formed.

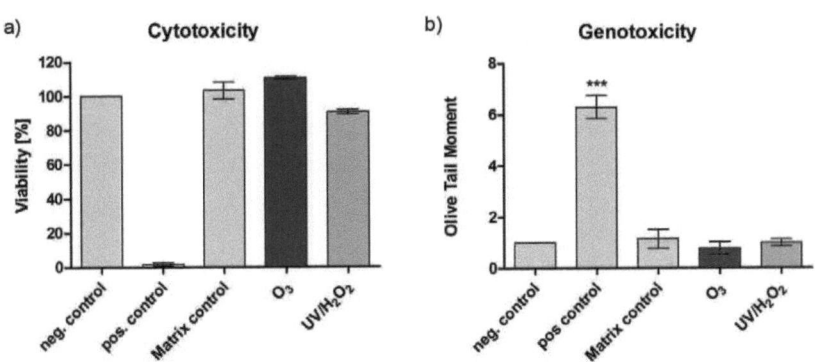

Figure 48: No cytotoxicity (a) or genotoxicity (b) of 18 mg/L Ofloxacin in WWTP effluent before (Matrix control) and after 60 min of ozonation (10 mg/L O_3) or UV/H_2O_2 treatment.

Results

5.5.4. Triclosan and 2,4-Dichlorophenol

Triclosan (2,4,4'-trichloro-2'hydroxydiphenylether) is an antimicrobial agent used in a variety of everyday life products, e.g. toothpaste, kitchen supplies, sports clothing, cosmetics, childrens toys etc. [170, 171]. Although it is not used as a pharmaceutical agent, its ubiquitous use has led to detections in water systems, including WWTP effluents and surface waters, in concentrations up to several µg/L where it might pose a threat to the environment [171, 172].

The ozonation of Triclosan using three different Triclosan:ozone ratios (1:1, 1:3 and 1:5) resulted in the formation of a variety of oxidation by-products with 2,4-Dichlorophenol being the most prevalent one as shown in Figure 49 [173].

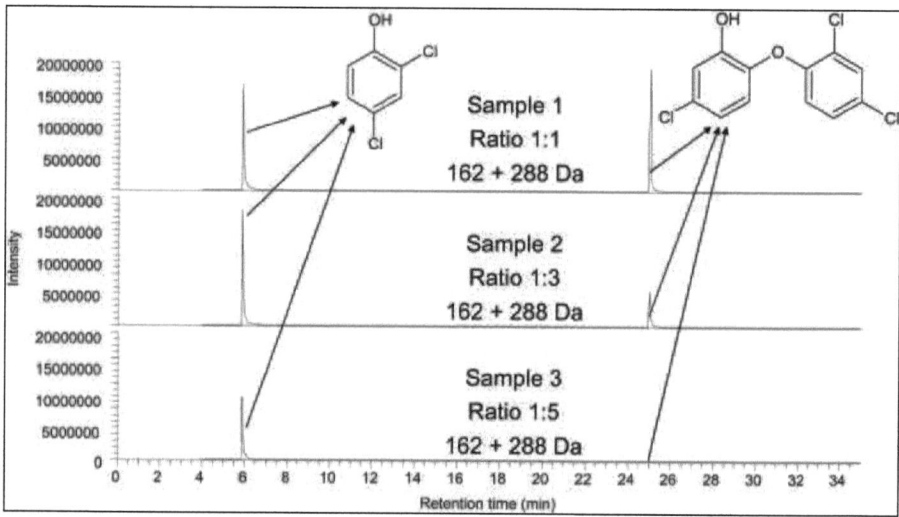

Figure 49: HPLC-MS/MS Q1-Scan chromatogram of a sample with different triclosan:ozone ratios of 1:1, 1:3 and 1:5 [173].

Therefore sterile concentration series of Triclosan and 2,4-Dichlorophenol (0 – 100 µg/L) were prepared in Ampuwa and the cytotoxicity as well as the genotoxicity were determined. The MTT test showed that neither substance had cytotoxic effects up to a concentration of 100 µg/L since the viability lies above 90 % for all tested concentrations (Figure 50a). In contrast the genotoxicity testing did reveal toxic effects. The higher concentrations of Triclosan result in a statistically significant increase in the Olive Tail Moment whereas the same concentrations of 2,4-Dichlorophenol did not result in significant changes compared to the negative control (Figure 50b). These results clearly show that the ozonation is a useful method in the removal of the genotoxic Triclosan.

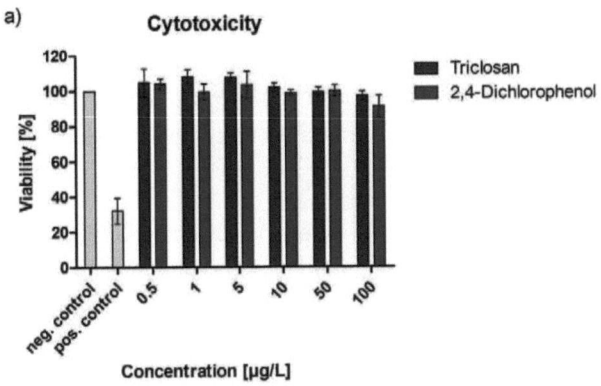

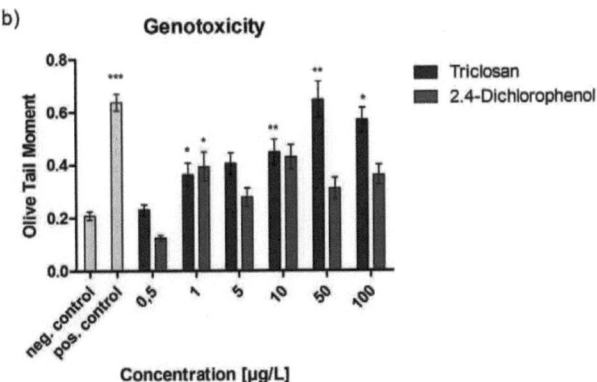

Figure 50: a) No cytotoxicity of 0 – 100 µg/L Triclosan or 2,4-Dichlorophenol. b) Genotoxic effects of triclosan but not 2,4-Dichlorophenol at higher concentrations.

Results

5.6. Biocides

According to the guideline 98/8/EG of the European parliament and the Council of the European Union, biocides are substances or compounds which are meant to destroy, render harmless or frighten off vermins as well as to prevent damages through them [174]. Since these substances are biological active they might pose a high environmental risk, including a risk to the human health. Due to their mode of action the use is widespread including agriculture, medicine and industry.

5.6.1. Irgarol 1051

Irgarol® 1051 also known as Cybutryn is used as an algicide e.g. on ship hulls or in antifouling paints e.g. on facades. According to its safety data sheet it is classified as being highly toxic to aquatic organisms [175].

7.5 mg/L Irgarol® 1051 were dissolved in HPLC-water and subjected to 56 or 400 µg O_3 for 60 minutes. The toxicity tests (MTT test and Alkaline Comet Assay) revealed no effects of 0.75 mg/L Irgarol® 1051. Neither cytotoxic (Figure 51a) nor genotoxic effects (Figure 51b) were detected since the viability as well as the Olive Tail Moment of the cells exposed to the samples are comparable to the values of the negative control. There is also no significant difference between the two tested O_3 concentrations in regard to the toxicity.

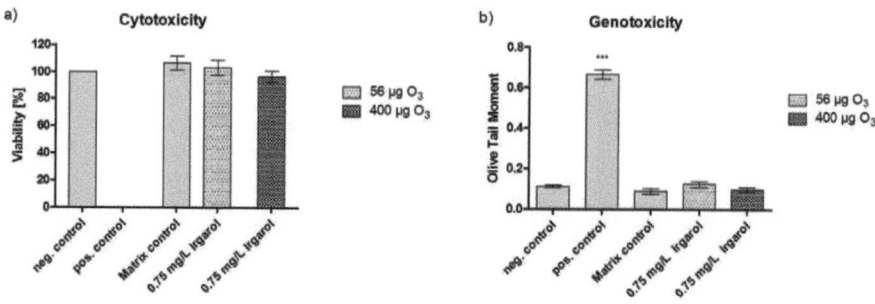

Figure 51: No cytotoxicity (a) or genotoxicity (b) of 0.75 mg/L Irgarol 1051 after ozonation.

Results

5.6.2. Terbutryn

Terbutryn is an herbicide and a constituent of a variety of products [176]. As Irgarol® 1051, Terbutryn works by inhibiting photosynthesis [177] and is thus used in agriculture as well as in the aquatic environment to prevent algal growth [176, 178].

The formation of possibly toxic oxidation by-products of Terbutryn has been studied using four different concentrations of ozone. Terbutryn was dissolved in HPLC-water and had a final concentration of 490 µg/L during the toxicity tests. The MTT test showed, that Terbutryn itself, as well as the samples after ozonation did not lead to effects on the mitochondria (Figure 52a), thus no cytotoxicity was detected. In contrast, the Alkaline Comet Assay revealed genotoxic effects. As can be seen in Figure 52b 490 µg/L Terbutryn already resulted in very significant (p = 0.001-0.01) DNA damages compared to the negative control. These effects were then increased after ozonation with 36, 100 and 195 µg ozone per liter resulting in highly significant (p < 0.001) DNA damages. In contrast, using an ozone concentration of 800 µg/L no genotoxic effects were observed.

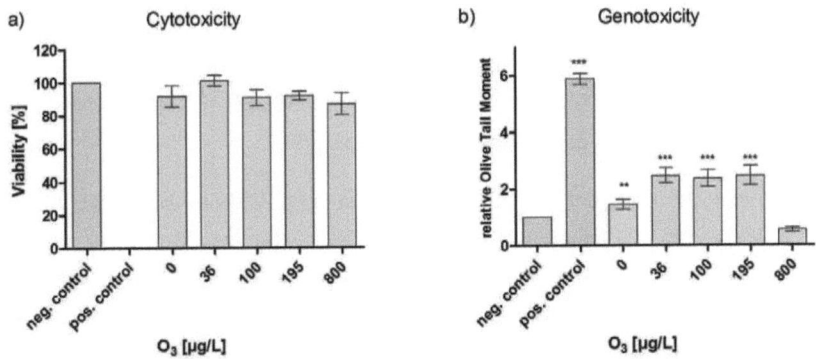

Figure 52: a) No cytotoxic effects of 490 µg/L Terbutryn in HPLC water before and after ozonation. b) Genotoxic effects of Terbutryn before and after ozonation (0 – 195 µg/L O_3) but no genotoxicity using 800 µg/L O_3.

Results

5.7. Musk fragrances

Musk fragrances are aromatic substances used in personal care products (e.g. shampoo, perfume, soap, etc.) with AHTN and HHCB being two of the most used compounds [179]. Their ubiquitous use (both classified as high production volume chemicals) has led to their detection in surface waters [38] and WWTP effluents since they are not removed during conventional treatment processes. Their presence in the water system is of great concern since musk xylene has already been added to the candidate list of substances of very high concern for authorization by the European Chemicals Agency (ECHA) [180]. In addition polycyclic musks, like AHTN and HHCB have also been detected in water and fish samples as well as in human fat [181].

5.7.1. AHTN

AHTN (8 1-(3,5,5,6,8,8-hexamethyl-6,7-dihydronaphthalen-2-yl)ethanone) also known as Tonalid is one of the most used polycyclic musks in cosmetics [179]. The exposure of AHTN on a daily basis was shown to be 409 µg/d after an application to the skin. Blood concentrations range up to a maximum of 0.29 µg/L [182] and environmental concentrations have been reported to be in the range of ng/L to µg/L concentrations [36, 37, 183, 184]. According to the SCHER risk assessment report, AHTN is not considered genotoxic or carcinogenic [185].

Up to 5 min UV treatment (low pressure UV-Hg lamp; 254 nm) of 1 mg/L AHTN in HPLC-water had no influence on the cytotoxicity, whereas 20 min UV treatment resulted in cytotoxic effects (Figure 53a) after an exposure to 0.1 mg/L AHTN. The viability of the used CHO cells was decreased to 74.3 % (± 7.7 %) thus the sample was classified as weakly cytotoxic. In addition, the cytotoxic sample was tested for remaining peroxides but none were detected indicating the formation of cytotoxic oxidation by-products since there was also no initial cytotoxicity of AHTN before the application of UV treatment. The Alkaline Comet Assay in contrast did not show any effects for all three time points of UV treatment (Figure 53b).

Results

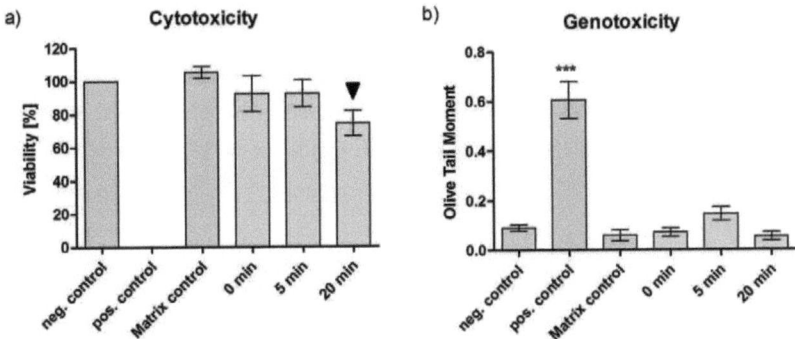

Figure 53: UV/H_2O_2 oxidation of 0.1 mg/L AHTN in HPLC water. a) Weak cytotoxic effects after 20 min UV treatment. B) No genotoxic effects before and after UV treatment. ▼cytotoxic effects

The same concentration of AHTN (0.1 mg/L) was also tested for toxicity before and after ozonation (10 mg/L O_3). Cytotoxic effects were neither detected before nor after 10 min or 60 min of ozonation. Looking at the results of the genotoxicity testing it can be seen that the Olive Tail Moment of all samples was not significantly increased compared to the negative control. Thus no genotoxic effects were detected after 10 and 60 minutes of ozonation.

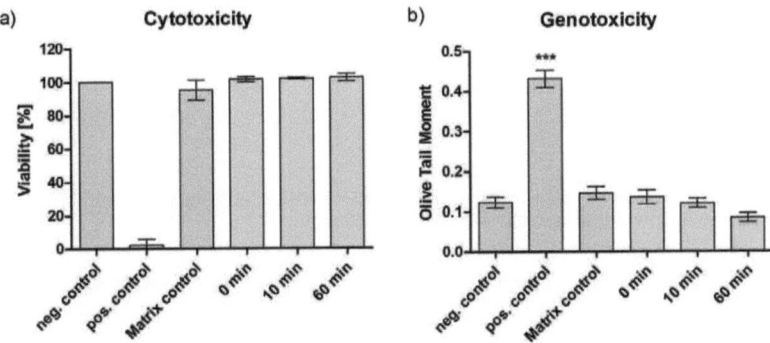

Figure 54: Ozonation (10 mg/L O_3) of 0.1 mg/L AHTN in HPLC-water. No cytotoxic (a) and genotoxic effects (b).

Results

5.7.2. HHCB

HHCB (4,6,6,7,8,8-hexamethyl-1,3,4,7-tetrahydrocyclopenta[g]isochromene) also known as Galaxolid® is a musk fragrance used in personal care products, e.g. shampoo or soap on a daily basis, thus it is detected in water systems. The use of cosmetics containing HHCB might result in an exposure of 51.2 µg/d by an uptake through the skin. It has also been shown to be present in the blood with the highest concentration being 6.9 µg/L [182].

The ozonation of HHCB resulted in the formation of the oxidation by-product HHCB-lactone [186]. Therefore concentration series (0.05 – 50 µg/L) of HHCB and HHCB-lactone were tested for toxicity. The MTT test revealed no cytotoxic effects for both substances of the tested concentrations (Figure 55a). The viability of all concentrations was above 90 % thus HHCB and HHCB-lactone are both classified as not cytotoxic up to a concentration of 50 µg/L. In addition there was also no genotoxicity detected for the same concentrations (Figure 55b).

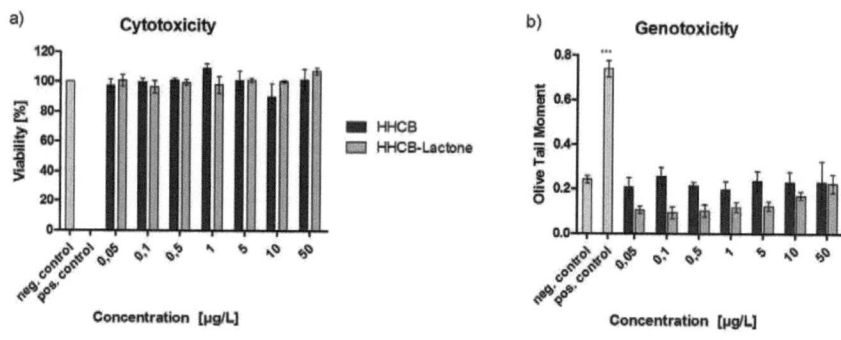

Figure 55: No cytotoxicity (a) and genotoxicity (b) of HHCB and its oxidation by-products HHCB Lactone at concentrations between 0.05 and 50 µg/L.

The UV/H_2O_2 oxidation of HHCB also resulted in the formation of oxidation by-products, however the chromatogram showed the formation of more and different by-products than the chromatogram of the ozone treated sample [186]. Toxicity tests of the untreated and the UV/H_2O_2 treated samples (15 W) of 0.1 mg/L HHCB in HPLC-

Results

water showed that only the sample after 60 min of UV treatment resulted in a decrease in viability to 54.2 % (± 0.7 %) viable CHO cells. In contrast, before treatment and after 30 min treatment no cytotoxic effects were detected (Figure 56a). The cytotoxic sample was also tested for remaining peroxides but none were detected, thus the effects can be related to formed oxidation by-products. Except the one cytotoxic sample (60 min), all others were tested for genotoxic effects using the Alkaline Comet Assay. Comparing the OTM of the untreated (0 min) and the treated sample (30 min) it can be seen, that the values were not significantly increased (Figure 56b), thus there was no genotoxic activity of 0.1 mg/L HHCB or the by-products at this time points of treatment.

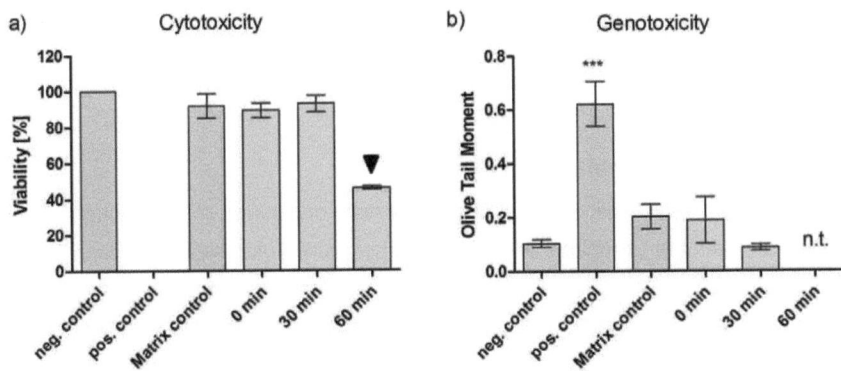

Figure 56: Cytotoxicity (a) and genotoxicity (b) of 0.1 mg/L HHCB in HPLC-water before (0 min) and after 30 min or 60 min of UV/H_2O_2 treatment (15 W). ▼cytotoxic effects; n.t. = not tested

Musk fragrances are also thought to exhibit estrogenic effects, however the ER Calux performed with the HHCB concentration series revealed no estrogenic activity (Figure 57). Comparing the RLU values of the HHCB samples and the 17ß-Ethinylestradiol standard series it can be seen, that they are lower than the negative control, thus smaller than the LOD and therefore the tested concentrations were not estrogenic.

Results

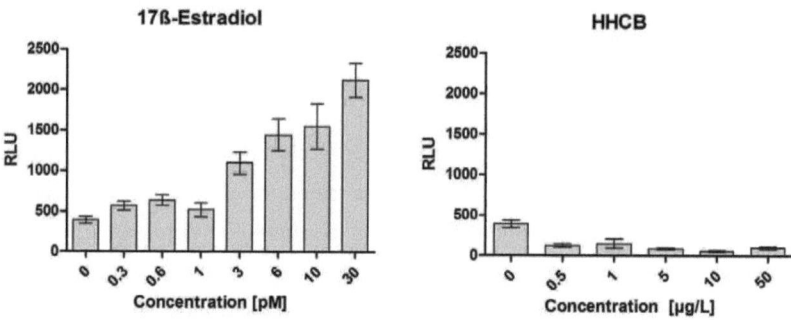

Figure 57: a) Relative light units (RLU) of the 17ß-Estradiol standard series. b) Relative light units of 0.5 – 50 µg/L HHCB.

5.8. Organophosphates

Organophosphate is the general term for esters or amides of the phosphoric acid. They are used as pesticides or additives in hydraulic fluid, plastic materials or flame retardants [37, 155].

5.8.1. Tris(2-chloro-1-methylethyl) phosphate (TCPP)

Like the other two organophosphates (TCEP and TPP) TCPP is used as a flame retardant in plastic materials.

1 mg/L TCPP were dissolved in HPLC-water and treated with ozone or UV/H_2O_2 for up to 60 minutes. Displayed in Figure 58a are the results of the MTT test. It can be seen that the ozonation has no effects on the toxicity of 0.1 mg/L TCPP. The UV/H_2O_2 treatment however resulted in a decrease of the viability to 68 %. Since there was no cytotoxicity detectable before treatment and the test for peroxides was negative, the effects can be related to formed oxidation by-products. The results of the genotoxicity testing of the ozonated and the UV/H_2O_2 treated sample at 0 min show no statistically significant increase in the OTM compared to the negative

control (Figure 58b). The cytotoxic sample (60 min UV/H$_2$O$_2$) was then not tested for genotoxicity.

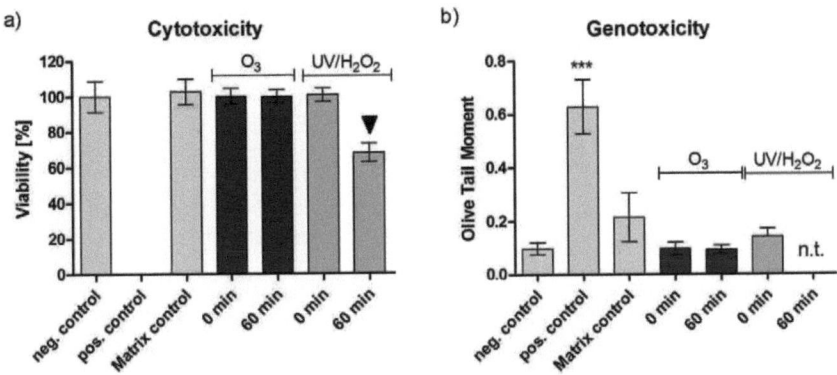

Figure 58: a) Cytotoxicity and b) genotoxicity of 0.1 mg/L TCPP before (0 min) and after (60 min) ozonation or UV/H$_2$O$_2$ treatment. ▼cytotoxic effects; n.t. = not tested

5.8.2. Tris(2-chloroethyl)phosphate (TCEP)

TCEP is also used as a plasticizer, flame-retardant and viscosity regulator for materials used in airplanes, vehicles or toys. Since the 1990's its use is declining due to the substitution with other substances e.g. TCPP [187].

No cytotoxicity was detected for 0.1 mg/L TCEP before and after 60 min of ozonation. The UV/H$_2$O$_2$ treatment (60 min) in contrast resulted in a loss of viability which was not due to remaining peroxides (Figure 59a). Thus toxic oxidation by-products have been formed. Except the one cytotoxic sample (UV/H$_2$O$_2$ 60 min) the other samples were also tested for genotoxicity (Figure 59b). No increase in DNA damage was detected, thus the samples are not genotoxic.

Results

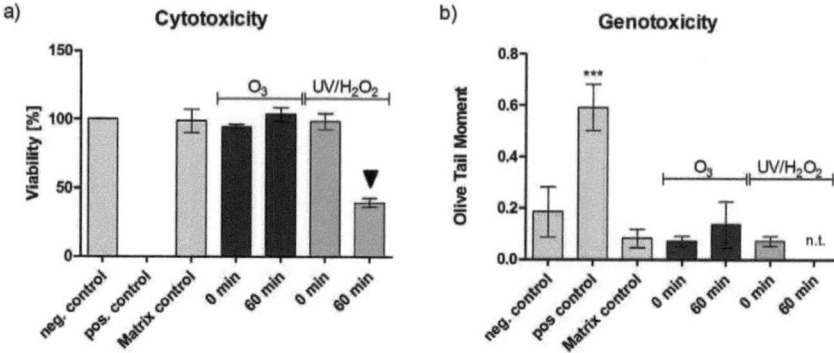

Figure 59: a) No cytotoxicity of untreated and ozonated TCEP in HPLC-water (0.1 mg/L), but cytotoxic effects after 60 min UV/H$_2$O$_2$ treatment. b) No genotoxic effects of non cytotoxic TCEP samples. ▼cytotoxic effects; n.t.=not tested

5.8.3. Triphenyl phosphate (TPP)

TPP is used as a flame retardant as well as a plasticizer which is, according to its safety data sheet, highly toxic to aquatic organisms and can have effects on the human health upon exposure.

TPP was dissolved in HPLC-water with a final concentration of 0.1 mg/L during exposure. Samples before and after ozonation (5 mg/L) or UV/H$_2$O$_2$ (Hg-LP 3 W; 1 mg/L H$_2$O$_2$) treatment were analyzed for cytotoxicity and genotoxicity. The results of the cytotoxicity test revealed, that neither sample resulted in a loss in viability (Figure 60a). However, the results of the Alkaline Comet Assay showed an increase in the OTM for the untreated samples whereas both the ozone and UV/H$_2$O$_2$ treated samples after 60 min resulted in no significant changes of the OTM (Figure 60b). The UV/H$_2$O$_2$ treated sample was in addition tested for remaining peroxides but none were detected. Thus the natural genotoxicity of 0.1 mg/L TPP was removed through the application of both oxidation methods.

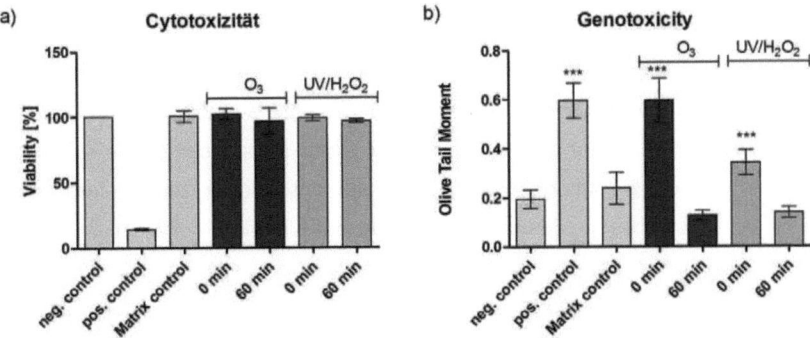

Figure 60: a) No cytotoxicity of 0.1 mg/L TPP in HPLC-water before and after ozonation or UV/H_2O_2 treatment. b) Genotoxicity only before oxidative treatment (0 min)

6. Discussion

Substances of anthropogenic origin are introduced into the water system by a variety of pathways. One major source of their introduction into surface waters is the effluent of waste water treatment plants. For most of these substances only a small part is removed during the common waste water treatment process. Since a large amount of those substances is designed to be biologically active they might have an influence on the ecosystem and even on the human health. Therefore additional treatment steps are needed in order to remove remaining micropollutants and at the same time preventing their release into the environment where they might cause harm. Despite the variety of possible removal methods, advanced oxidation processes are the most promising techniques due to their low costs, ease of applicability and efficiency. However one drawback of these processes is the formation of oxidation by-products through an incomplete degradation. These by-products might then differ in their toxicological properties. Due to a lack of information on the toxicity of many micropollutants and their by-products a general statement on the safe use of Advanced Oxidation Processes is difficult.

The efficiency of the here used AOP methods (Ozonation and UV/H_2O_2 oxidation) was evaluated based on the formation of oxidation by-products and their toxicity. Therefore pure water as well as waste water treatment plant effluent samples were spiked with pharmaceuticals, musk fragrances, estrogenic substances, biocides or organophosphates and subjected to oxidation. Samples were taken time dependently starting before the oxidation. Afterwards their ability to induce cytotoxic and genotoxic as well as to some part estrogenic and mutagenic effects was examined.

Discussion

6.1. Endocrine disruption

Although the focus of this work in regard to endocrine disruption was laid on estrogenicity, other effects like androgenicity, glucocorticoid activity or thyroid hormone like activity of water samples etc. can not be excluded. Different classes of chemicals (phthalates, polychlorinated biphenyls, pesticides etc.), either of natural or synthetic origin, detected in water systems have already been identified to be endocrine disrupting because of their structural similarity compared to hormones [188-192]. In 2000 the BKH Consulting Engineers published a list of priority substances based on existing lists resulting in 564 substances which are suspected to be endocrine disrupting. Out of this list 146 chemicals were evaluated and 66 were classified as "Category 1" chemicals indicating that there is certain evidence for endocrine disrupting effects in living organisms. Most of these Category 1 chemicals were also considered having a high exposure concern [193]. Outcomes of an exposure to endocrine disruptors, e.g. the feminization of alligators, birds or fish have been linked to the presence of endocrine disruptors [194-198]. Most of the endocrine effects detected so far are affecting the estrogenic cycle. It has been shown that e.g. obesity or certain kinds of cancer besides other health effects are related to an exposure to estrogens [199-204]. This is supported by the US EPA stating that endocrine disruption is a mode of action which potentially might lead to adverse effects e.g. carcinogenic, developmental or reproductive effects. However they also state, that only a limited number of studies are available showing a direct correlation between estrogen exposure and adverse effects in humans especially in the development of cancer [191].

It is now questionable whether humans who are exposed to endocrine disrupting chemicals in drinking water, food, etc. are susceptible to hormonal disturbances thus adverse health effects. Even though there is no direct linkage between endocrine disruptors in the environment and human health impacts it can not be neglected. Animal data clearly proves a correlation between biological active substances and endocrine disruption. Due to the high persistence of these substances long-term studies investigating single compounds as well as complex mixtures are needed for a risk assessment. Long-term studies covering chronic exposures are also important in regard to low-dose effects (concentrations below the physiological level) of single substances and complex substance mixtures because this is the most probable

Discussion

scenario of exposure in the environment. The term low-dose effects has been defined as those biological effects which occur at the range of a typical human exposure or a dose lower than those which are typically used in the US EPA standard testing paradigm [205]. Although endocrine disrupting effects are usually detected at concentrations higher than those found in the environment, knowledge about a chronic exposure to low concentrations is scarce. In a review by Vandenberg et al. (2012) it is concluded that results from a high-dose exposure study can not be used to predict effects of a low-dose exposure [206]. In addition combination effects such as additive or synergistic effects occurring in complex water matrices through the presence of a variety of substances with the same mode of action need to be considered when discussing low-dose effects (see chapter 6.3 on mixture effects). Low-dose effects also might differ from effects detected at higher concentrations. Reasons for this might be cytotoxic effects as well as a down-regulation of the receptors of high concentrations whereas low-doses might result in an increased receptor acitivity thus an up-regulation of specific genes [206]. Therefore measures should be taken not only to detect and identify endocrine disrupting micropollutants in water samples but also to investigate their possible health effects at high and low doses and finally to develop methods to successfully remove them from the environment.

Since the effects of hormones occur at very low concentrations it is necessary to use test systems which are able to detect endocrine disruptors at those low concentrations. Kase et al. (2009) therefore reviewed available test systems applied in the aquatic environment for the detection of endocrine disrupting substances. A huge variety of *in vitro* test systems is available and examples as well as their mode of action are displayed in Figure 61. This battery of tests then allows a mode of action based risk assessment which can also be modified according to certain needs. They conclude that a combination of *in vivo* and *in vitro* methods is needed to gather information on the different levels, namely the hormone activity and endocrine disruption as well as reproductive effects [71].

Discussion

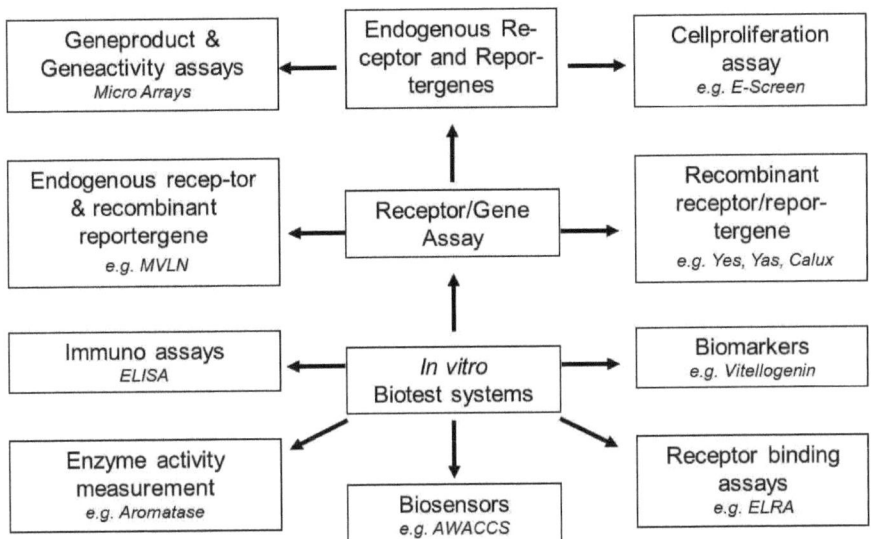

Figure 61: *In vitro* test methods for the determination of endocrine effects adapted from Kase et al. 2009 [71]

Different other publications also show the applicability of *in vitro* test systems as sensitive methods for the determination of endocrine effects. However they also conclude that a risk assessment should not only be based on *in vitro* methods, thus a combination of *in vitro* and *in vivo* methods is favorable [207-210].

6.2. Antibacterial activity during advanced oxidation processes

Another concern of antibiotics and their transformation products present in water systems is their potential to support the establishment of resistances [3]. The isolation and identification of bacterial strains exhibiting certain resistances is steadily increasing. Due to the widespread use (human and veterinary) and the high amount of antibiotic prescriptions annually they are ubiquitously found in the environment, including groundwater, at concentrations up to several µg/L [3, 211]. Resistances are distinguished by either being primary (naturally occurring, without gene mutation) or secondary (acquired during exposure through e.g. mutations or

Discussion

gene transfer) [212] with the latter one being the interesting one in regard to water systems. The importance of this problem, e.g. the ineffectiveness of antibiotic therapy especially for the treatment of nosocomial infections, is well understood and in 2001 the WHO released a Global Strategy for Containment of Antimicrobial Resistance [213]. In addition the German Ministry of Health also published a strategy for further actions regarding antimicrobial resistances [212]. One of the most prominent cases of antibiotic resistance is the Methicillin-resistent *Staphylococcus aureus* (MRSA) strain, first detected in the early 1960's against which most of the antibiotics are not effective [214, 215]. While *Staphylococcus epidermidis* is the nonpathogenic form found on human skin and mucuous membranes, *Staphylococcus aureus* is the pathogenic form associated with e.g. pneumonia or meningitis [216].

Because of these resistances advanced oxidation processes should also take into account the complete removal of antibacterial activity resulting from antibiotics present in waste waters. Therefore it is also important that formed oxidation by-products also do not exhibit antibacterial acitivty. A variety of studies investigating the antibacterial activity during oxidation processes were already published. Dodd et al. (2009) were able to prove the loss of antibacterial activity due to oxidation for e.g. Sulfamethoxazole, Ciprofloxacin and Triclosan whereas the oxidation of Penecillin G and Cephalexin led to the formation of by-products with antibacterial activity [217]. As a part of the here presented project Claudia vom Eyser (IUTA, Duisburg) therefore also looked at the bacterial growth inhibition by water samples (pure water and WWTP effluent) spiked with Ciprofloxacin or Ofloxacin in her Master Thesis (2011). She found that the antibacterial effect decreased over time of oxidative treatment thus the formed oxidation by-products have no antibacterial effects [218]. Similar results have been shown by Wammer et al. (2006). They tested three sulfa drugs (Sulfathiazole, Sulfamethoxazole, Sulfachloropyridazine) and Triclosan as well as their transformation products after photochemical treatment for their potential to inhibit bacterial growth. Although a growth inhibition was detected for the parent compound none of these tested antibiotics revealed antibacterial acitivities after photolysis [17].

Discussion

6.3. Mixture effects

Although the here presented results as well as published data demonstrate that adverse toxicological effects of single substances mostly occur at much higher concentrations than those usually detected in waste water treatment plant effluents and surface waters, mixture effects need to be considered. This becomes even more apparent since not each single micropollutant as well as their metabolites and transformation products can be determined or investigated regarding its toxicity. Organisms are usually not solely exposed to single substances but rather to complex mixtures containing a variety of substances with different toxic effetcs. In addition the amount of TOC (total organic carbon) does not give any information on the biological activity especially the toxicity and therefore whole effluent samples should be analyzed. A variety of studies have been performed proving a good accuracy of concentration addition models for substances with a similar mode of action [219-221]. In this context e.g. Chèvre et al. (2006) propose the use of a risk quotient for the assessment of herbicides with a similar mode of action based on a concentration addition model rather than using a limit value of 0.1 µg/L for each individual pesticide as a water qualitiy criteria in Switzerland [222]. In Germany the drinking water guideline also sets a limit value of 0.1 µg/L for individual biocides and their metabolites as well as a limit value of 0.5 µg/L for the sum of pesticides [85]. Brian et al. (2005) investigated the estrogenic effects of Estradiol, Ethinyestradiol, Nonylphenol, Octylphenol, and Bisphenol A individually for each substance as well as for mixtures. They found that even a concentration below the effect level of each single substance will result in effects after mixing with a good correlation to a concentration addition model [223]. But it has also to be noted that a concentration addition model can only be applied when the constituents of a mixture as well as their concentrations are known. In contrast, the effect prediction of mixtures containing substances from different classes and different modes of action are not that easily estimated by concentration addition or response addition since synergistic, antagonistic or additive effects might occur (Table 31) [224-226]. Hernando et al. (2003) were able to show that the toxicity of pesticides increased in the presence of Methyl-tert-butyl-ether, a fuel oxygenate also commonly found in water systems [227]. In another study with different pesticides synergistic, antagonistic as well as additive effects of substance mixtures on different organisms

Discussion

were seen. The results also indicated that the mixture effects were not always the same in different aquatic organisms [228].

However, despite these studies, knowledge about mixture effects is still scarce especially for water environments and needs to be further investigated for a secure risk assessment.

Table 31: Possible mixture effects in regard to biological effects

Effect	Description
Synergistic	Substances acting together resulting in an increased effect, thus an effect higher than would have been predicted by the addition of the potency of the single compounds
Antagonistic	Inhibition of a substance by another substance resulting in a decreased effect
Additive	All substances of a mixture with a common mode of action resulting in a jointly action of effect addition, by proportionally contributing to the overall effect even at a concentration lower than their observed effect level

To overcome this gap of knowledge on mixture effects, methods like the QSAR approach (Quantitative Structure-Activity Relationship) or the use of –omics techniques might be a solution. Modelling the effects of single substances using QSAR approaches can then be used for an assessment of mixture toxicity.

Discussion

6.4. Use of toxicological *in vitro* methods for water quality control

Most regulations nowadays still require the use of animals for risk assessments. However, the vast amount of compounds (existing and new ones) which need to be analyzed by means of toxicity is far too huge to be tested *in vivo*. Animal testing requires a large amount of test animals and in addition it is cost and time consuming. Another fact to consider is the extrapolation of results gained from *in vivo* studies e.g. aquatic or rodent studies to a human health risk assessment. Due to inter- or intraspecies effects this extrapolation is not always possible and a certain risk remains. To overcome this gap uncertainty factors also called safety factors have been used to set values for an acceptable daily intake. These take into account the intra- as well as the interspecies variation and the quality of the data (e.g. short-term, long-term, study size) [229, 230]. Although uncertainty factors can be applied, there still might remain a possible risk for humans.

To fully understand the whole range of possible health effects through emerging contaminants, an approach at the cellular level for a screening purpose is promising. Cellular test methods can be applied as high-throughput systems with low costs. An improvement in this matter are reporter gene assays, e.g. Calux test systems, which are more sensitive than e.g. yeast cell based assays [71]. The use of the so called "–omics" techniques might be another step towards the application of cellular based test systems. These techniques allow an analysis on the whole organism situation e.g. changes in DNA expression (genomics) or protein profiles (proteomics) [231]. For further or more detailed analyses *in vivo* methods might then be applied after limiting the vast amount of adverse effects through *in vitro* testing.

Although the use of toxicological *in vitro* methods is not a general parameter in most guidelines for water quality testing, they have been proven as a relevant enhancement in this matter [22, 26, 232, 233]. Chemical methods are limited to a qualitative and quantitative detection of micropollutants, but a statement about their biological effects (e.g. toxic, bactericidal) is not possible. In this context a combined approach of chemical and biological analyses, called effect-directed analysis (EDA) combining chemical analyses, biological test systems and fractionation processes for a cause-effect relationship is most promising [234, 235] (Figure 62).

Discussion

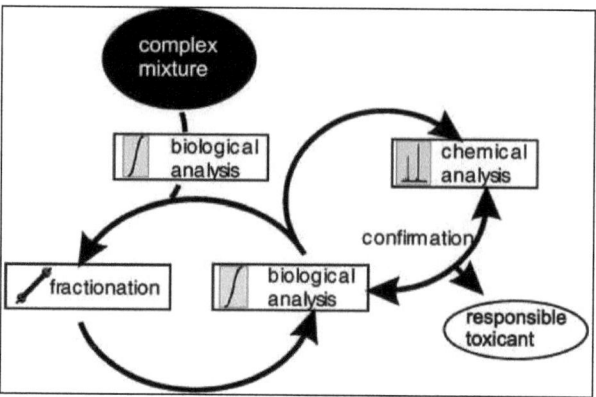

Figure 62: Schematic overview of the EDA (effect directed analysis) of mixtures published by Werner Brack [234].

An approach called "Adverse Outcome Pathway" published by the US EPA (2010) (Figure 63) also suggests the use of cellular and molecular techniques as a basis (Anchor 1) followed by tests at the organism- and population-level (Anchor 2) combining both test principles to link an effect at the molecular level to an adverse outcome at the organism/population level, spanning different biological levels [236].

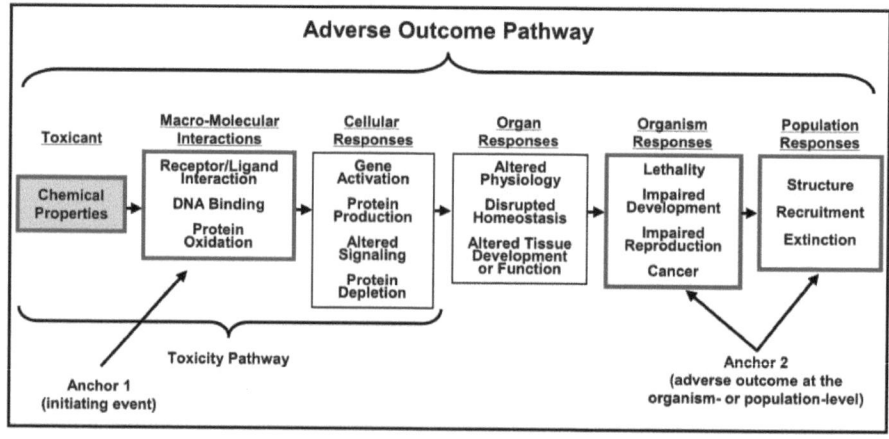

Figure 63: Adverse Outcome Pathway proposed by the US EPA for risk assessment based on *in vitro* and *in vivo* methods [236].

Discussion

According to a variety of studies bioassays in general but especially cell culture based systems have been established world wide for an assessment of the risk resulting from organic micropollutants as well as their by-products (oxidation and disinfection) [22, 26, 158, 237]. The variety of *in vitro* methods covering different toxicological endpoints is large and steadily growing and improving. With the use of non specific toxicity tests e.g. cell proliferation assays a general statement about the overall toxicity of a sample is possible including mixture effects. The additional use of more specific *in vitro* tests investigating reporter gene activities, DNA damaging potentials etc. will then help to establish a human health risk assessment.

A combination of *in vitro* and *in vivo* methods allows a screening at the molecular level through cell based systems as well as a more targeted analysis using *in vivo* methods. The addition of chemical analyses and aquatic studies will then give a complete overview starting with the detection of a substance, the determination of the concentration, the detection of molecular and systemic effects of a certain substance or mixture effects, e.g. dose-response curves, or effects on whole populations.

6.5. Applicability of ozonation and UV/H_2O_2 oxidation during waste water treatment

Advanced oxidation processes function through the formation of highly reactive hydroxyl radicals which then attack the substituents of the organic matrix (e.g. through electron transfer, hydrogen abstraction, etc.) and initiate the mineralization of these compounds [238]. Despite the advantages of those methods the formation of oxidation by-products needs to be considered since a complete degradation might not be economical. Therefore the efficiency of the applied methods is dependent on the ease of applicability, the cost effectiveness and the potential to remove as much DOC (dissolved organic carbon) as possible without generating bioactive, e.g. endocrine disrupting, genotoxic or mutagenic by-products [53, 239]. In addition to oxidative techniques, sand filtration or activated carbon addition might be applied as

Discussion

the last treatment step to yield a higher degree of micropollutant and especially metabolite and oxidation by-product removal [14, 24, 240, 241].

In regard to the efficiency of the two in this study investigated advanced oxidation processes (UV/H_2O_2 and Ozonation), the results of the toxicological testing presented here demonstrate that these methods are applicable for waste water treatment plant effluent matrices. Neither cytotoxic nor genotoxic effects were detected before and after the oxidative treatment of the two ß-blocker Atenolol and Metoprolol, the antibiotics Ciprofloxacin, Ofloxacin and Sulfamethoxazole as well as the musk fragrance AHTN and the biocide Irgarol 1051 at the tested concentrations. In addition Ciprofloxacin was also found not to be mutagenic in the Ames test using the *Salmonella typhimurium* strains TA98 and TA100.

The results also indicate that their oxidation by-products, formed as a result of an incomplete degradation, are also not toxic and the UV/H_2O_2 treatment as well as the ozonation are sufficient methods for the successful removal of these micropollutants from the effluents of waste water treatment plants using the above mentioned settings. Although other studies showed toxic effects of these substances the concentrations used were to some extent much higher than the concentrations used in this study and the concentrations detected in water bodies.

The oxidation of HHCB, Ciprofloxacin, and Bisphenol A clearly showed that the structure of the formed oxidation by-products as well as their toxicity might be dependent on the used method as well as the water matrix which is supported by other published studies [16, 20, 66, 242].

Overall these results clearly demonstrate that the treatment duration, the used amount of oxidant as well as the oxidation method are of great importance in regard to a complete removal of the substances and the loss of biological (antibacterial, toxic, endocrine etc.) activity. It has also been shown, that the here used *in vitro* methods are applicable for the testing of water samples which is supported by other previously published studies, proving the usefulness and applicability of a variety of *in vitro* test systems [14, 22, 26, 232, 243].

These results are in accordance with other published studies, proving the applicability of oxidative methods for water purification. It has been long known that

Discussion
___

oxidative methods are useful in the removal of organic micropollutants. However, it is also known that an incomplete oxidation might result in the formation of oxidation by-products with different toxicological profiles and therefore should be prevented and measures be taken [241, 244-246].

6.6. Concentration of water samples

The steady improvement of chemical analytical methods is one reason for the increase in the amount of detected organic micropollutants over the last years. The concentration of water samples and high resolution analytical tools allow the detection of substances as low as ng/L concentrations. In addition to the benefit of a lower detection limit disturbing substances e.g. phosphate, nitrate or ammonia are also removed during concentration. A disadvantage of the concentration step before a biological test might be the loss of substances. Therefore the extraction method should take into account a yield of as much substance as possible and at the same time an exclusion of as many interfering substances as possible. Desbrow et al. (1998) were able to show that the concentration of an effluent sample using a C18 column resulted in a loss of estrogenic activity of 80 % depending on the elution method [247]. Erger et al. (2012) investigated the influence of water residuals on the SPE performance and subsequent chemical analysis and found far-reaching influences on the performance of the GC-MS analysis [248]. Thus extraction methods should be thoroughly evaluated (e.g. using an internal control) to e.g. prevent the loss of substances.

Another advantage of extraction methods is that they can also be used to improve the sensitivity of biological test systems as well as analytical tools and to identify substances with a low concentration in water samples. This is especially important in the case of bioactive substances, e.g. those mimicking the hormonal activity, because they are usually active at very low concentrations. Since environmental samples contain a complex matrix extraction methods should also be capable of extracting several pollutants at the same time.

Discussion

A further approach in regard to the monitoring and detection efficiencies is the method of passive sampling. The principle behind this method, independent of the used system, is the free analyte flow from the tested medium (e.g. waste water treatment plant effluent) to the sampler material which due to different chemical gradients is a result of varying e.g. concentrations or pressures [249, 250]. An advantage of passive sampling is the identification of the average concentration of the analytes over a certain period of time, called time-weighted analysis [250]. A few studies as well as reviews already showed the use of passive samplers in environmental analyses, especially with water systems. They are for example used during monitoring processes of micropollutants as a screening tool [249, 251]. Their use in research over the past years is increasing and has been shown as a good alternative to active sampling [252-255]. Since each method has its advantages and its disadvantages there is no single method for all applications. Filtration steps before an SPE might result in the loss of substance whereas liquid-liquid extractions require organic solvents which are toxic and the whole procedure is time consuming. Thus depending on the purpose of the study each method should be evaluated and optimised before analyses.

6.7. ß-blocker

Due to their longterm use and high amount of ß-blocker intake they are commonly detected in water systems and might pose a threat to the environment. Concentrations of Metoprolol in waste water treatment plant effluents range up to almost 4 µg/L [256]. Therefore two of the most prescribed ß-blocker, namely Metoprolol and Atenolol, were investigated.

In the here presented study no cyto- or genotoxicity was detected for Atenolol and Metoprolol (chapter 5.3) before and after ozonation or UV/H_2O_2 oxidation. There were also no differences observed between pure water and waste water treatment plant effluent spiked with either of the ß-blocker. These results reflect what has already been published by others. Using *Ceriodaphnia dubia* (*C. dubia*) an EC_{50} of 33.4 mg/L for Atenolol and 45.3 mg/L for Metoprolol was determined by Fraysse et

Discussion

al. (2005) [257]. An acute toxicity study using the same organism also revealed that the EC_{50} values range from 1.4 to 163.4 mg/L after two days of exposure depending on the tested ß-blocker (Acebutolol, Atenolol, Metoprolol, Nadolol, Oxprenolol, Propranolol) [257]. Cheong et al. (2008) used the MTT test to detect cytotoxic effects of ß-blockers, including Atenolol and Metoprolol, on human corneal epithelial and retinal cells. They also found great differences in IC_{50} values when comparing eight ß-blockers. Atenolol resulted in an IC_{50} of 6.13 g/L and the IC_{50} for Metoprolol was 2.74 g/L [258].

Another study showed that no genotoxic effects occurred after a chronic exposure to 50 mg Atenolol per day [259]. In addition, tests with different cell types (haematocytes, gill and digestive gland primary cells) of the zebra mussel did not reveal notable toxic effects [260]. Ioannou et al. (2011) were able to show that the solar/TiO_2 degradation of 10 mg/L Atenolol first resulted in an increased mortality of *D. magna* which then decreased over time thus they concluded that the formation of toxic oxidation by-products led to these effects [261].

Huggett et al. (2002) also tested the toxicity of Metoprolol and found 48 h LC_{50} values of 63.9 mg/L in *D. magna* and 8.8 mg/L in *C. dubia* whereas the LC_{50} values were above 100 mg/L when using *H. azteca* and Medaka as test organisms [262] and the EC_{50} in *Danio rerio* was determined as 31 mg/L [263]. Tests with *Daphnia magna* resulted in an EC_{50} value of 200 mg/L [264] or an LOEC of 12 mg/L after a 9-day exposure [265]. Thus great species differences can be seen

Compared to the concentrations used in this study (Atenolol = 0.2 mg/L; Metoprolol = 0.1 and 1.4 mg/L) toxic effects are detected at much higher concentrations. The amounts of ß-blockers commonly found in waste water treatment plant effluents and surface waters are usually in the µg/L range or even lower [2, 8, 266]. Benner et al. (2009) were able to demonstrate the formation of oxidation by-products as well as reaction pathways of Metoprolol after ozonation [267]. A few of these ONP have been confirmed by our own studies but new ones were also found [151]. However, although by-products were found after oxidative treatment they did not have an effect on CHO cells in the here used test systems. As has been shown by own studies as well as published data by other groups toxic effects occur at concentrations much higher than commonly found in water bodies.

Discussion

6.8. Estrogenic substances

6.8.1. Ethinylestradiol

As a main constituent of contraceptive pills Ethinylestradiol is excreted and found in waste water treatment plant effluents as well as other water systems at ng/L concentrations [4, 268-270] and might therefore have an influence on e.g. the reproduction of aquatic organisms. Therefore these substances should be completely eliminated during waste water treatment before the effluent reaches surface waters.

Own results show that there was no cytotoxicity detected for Ethinylestradiol but estrogenic effects have been detected before and throughout the entire duration of oxidative treatment, however slightly diminished because only a minor decrease was seen over time. A variety of studies have been published concerning the toxicity, including the estrogenicity of Ethinylestradiol. No mutagenic effects were detected using the chromosomal aberration test in human lymphocytes and mouse bone marrow cells [271] and the Ames test [272]. However, Siddique et al. (2005) were able to show genotoxic effects of 5 µM (~ 1.5 mg) Ethinylestradiol. These effects, in fact, were only detected after metabolic activation and the addition of NADP [273]. Estrogenic effects were, amongst others, detected by Dussault et al. (2008) showing an influence on the reproduction starting at 0.36 mg/L Ethinylestradiol [269]. Medaka exposed to 0.06 mg/L Ethinylestradiol revealed effects on their fertility [274] and Salierno et al. (2009) detected behavioral changes as well as changes of the hormonal system and the sex characteristics in fish exposed to 0.04 mg/L Ethinylestradiol [275]. Thus comparing the above results to the concentration used in this study, a possible impact on the environment might occur after the release of Ethinylestradiol into surface waters. This is also supported by a study published by Caldwell at al. (2008). They reviewed the available literature on the aquatic toxicity of 17α-Ethinylestradiol to derive a PNEC (predicted no effect concentration) value and concluded that a PNEC of 0.35 ng/L could be recommended as a safe measure for water systems, even for chronic exposures. They also concluded, that even a concentration of 0.5 ng/L might have no effects on the ecosystem [276]. However this value does not take into account e.g. additive effects resulting from the complex matrix of a water sample as well as low dose and long term effects (see chapter 6.1).

Discussion

Comparing this PNEC value to the results from this study it can be seen, that the here used ozonation and UV oxidation did not lead to a decrease of Ethinyestradiol below 0.35 ng/L thus in this case both treatment methods were not sufficient in removing the biological activity.

6.8.2. Bisphenol A

As a basic material during polycarbonate synthesis, Bisphenol A is found in many products of the everyday life e.g. reusable water bottles and other food containers. Trace amounts of it might then leach into foods and drinks. Because of its health impacts the EU enacted a directive (2011/8/EU) which banned Bisphenol A from the production of infant feeding bottles made of plastics starting in March 2011 [277].

Own toxicity tests with Bisphenol A performed in the here presented study indicated that toxic effects are linked to the used water matrix (chapter 5.4.2). The oxidation of waste water treatment plant effluent containing Bisphenol A did not result in the formation of toxic oxidation by-products. In addition, no effects were detected before and after ozonation of pure water containing 1.4 mg/L Bisphenol A whereas cytotoxic effects were seen after the UV oxidation of 0.1 mg/L Bisphenol A in pure water. These effects can be related to formed oxidation by-products since no remaining peroxides were detected in the sample. Comparing these results to published data it can be seen that differences in toxic effects between pure water and waste water treatment plant effluent due to different reaction pathways and competing reactions have already been described before [16]. When comparing distilled and surface water a better degradation of Clofibric acid and Ibuprofen was observed in distilled water [20]. Trovo et al. (2009) were able to demonstrate that the Photo-Fenton treatment of distilled water and seawater containing Sulfamethoxazole only resulted in an increase in toxicity when oxidatively treating seawater [66]. Differences in regard to toxic effects of substances were also seen when dissolved in seawater or distilled water, thus again the water matrix (salinity, amount of other sustances) had an effect on the structure of the formed by-products and thus the toxicity [242].

Apart from the cytotoxicity and genotoxicity testing the estrogenic activity of Bisphenol A before and after UV/H_2O_2 oxidation was also examined as a part of the

Discussion

here presented study. The results displayed in chapter 5.4.2 show that the oxidation was sufficient in removing the estrogenic activity of 0.1 mg/L Bisphenol A from HPLC-water. While there were estrogenic effects observable before UV/H_2O_2 treatment, there were no more estrogenic effects detected after 10 and 60 min of treatment.

Toxic or in general biological effects of Bisphenol A described in the literature are mainly based on its estrogenic effects [278]. As already shown for other substances, EC_{50} values vary (10 – 253 µg/L) between organisms, test methods and the duration of exposure used [279, 280]. Golub et al. (2010) reviewed the available literature in regard to toxic effects of Bisphenol A and found that it mainly affects the offspring viability, the sex-differentiation, the immune hyperresponsiveness and the gender-differentiated morphology, with the specific endpoint being dose related [281]. In addition to the already published data, Bisphenol A is one of 90 substances listed in the Community Rolling Action Plan (CoRAP) released by the ECHA in February 2012. The CoRAP includes substances which are suspected to have a risk on human health and therefore need to be further evaluated within the next three years. Bisphenol A was added to this list because of its potential endocrine disruptive activity as well as its widespread use [282, 283].

The results for Ethinylestradiol and Bisphenol A from our study demonstrate that there are differences in their degradation according to the detection of estrogenic effects. UV/H_2O_2 treatment of Bisphenol A resulted in a complete removal of estrogenic effects. In contrast, neither the UV/H_2O_2 treatment for up to 60 min, nor the use of up to 10 mg/L ozone resulted in a complete removal of the estrogenic activity of Ethinylestradiol after 60 min of treatment. Schrank et al. (2009) proved that the removal efficiency of estrogenic effects is strongly related to treatment parameters, especially the pH. Low pH values resulted in an increase in estrogenicity while a high pH during treatment resulted in a complete removal of estrogenic effects. This might be due to the fact that a decomposition of ozone takes place at higher pH values leading to the formation of radicals [284]. Schrank et al. (2009) also showed that despite a decrease in acute toxicity towards *V. fisheri* and *D. magna*, estrogenic effects occur after ozonation even though the untreated sample did not exhibit estrogenic effects [244]. A reason for this might be the

Discussion

elimination of substances which are structurally similar to estrogens and have a higher binding affinity thus blocking the estrogen receptor and preventing estrogenic substances from binding to it. Once these antagonistic substances are removed from the sample through ozonation estrogenic substances are then able to bind to the receptor and initiate DNA transcription. Another fact to consider in regard to estrogenicity and endocrine disruption in general is the exposure to low doses over long periods of time (see chapter 6.1 on low-dose effects).

6.9. Antibiotics

Antibiotics are pharmaceuticals ubiquitously used for the treatment of a great variety of infections. Depending on their mode of action and point of attack they are classified into different groups. Antibiotics are biologically active and might therefore pose a threat to the environment. Concentrations commonly detected in waste water treatment plant effluents are in the µg/L range. In this study one antibiotic from the group of sulfonamides (Sulfamethoxazole) and two fluoroquinolones (Ciprofloxacin and Ofloxacin) as well as the bactericide Triclosan have been investigated.

6.9.1. Sulfamethoxazole

Sulfamethoxazole belongs to the group of sulfonamide antibiotics which were originally used as bactericides. Nowadays some diuretics are also based on the sulfonamide group. The widespread use of sulfonamides has led to their detection in water bodies with concentrations up to 2 µg/L [285]. Therefore techniques for a complete removal of these substances are needed. Sulfamethoxazole was chosen as a representative of this group of antibiotics to investigate the effects of oxidative treatment on its toxic properties.

The above presented results for Sulfamethoxazole (5.5.1) indicate that neither the ozonation nor the UV/H_2O_2 oxidation resulted in toxic effects. Although the formation of oxidation by-products has been described [151, 286-288] none of the treated samples had an effect. This is in agreement with the results published by Yargeau et al. (2008) who tested a mixture of Sulfamethoxazole degradation products on human

Discussion

hepatocellular carcinoma cells (HepG2) and found no toxic effects on cell morphology or metabolism [288]. Another study showed that using TiO_2, the photocatalysis of Sulfamtheoxazole also leads to the formation of by-products which show a lesser toxicity on the green alga *Chlorella vulgaris* than the parent compound [289]. In contrast Dirany et al. (2011) detected toxic effects of intermediates formed during the electro-Fenton treatment of Sulfamtheoxazol in the *V. fischeri* luminescence test and thus an increase in toxicity compared to the untreated solution [245]. Toxic effects were also reported by del Mar Gomez-Ramos et al. (2011) after the ozonation of Sulfamethoxazole using *Daphnia magna* and *Pseudokirchneriella subcapitata* as test organisms proving the formation of toxic oxidation by-products [290]. Dantas et al. (2008) used the *V. fischeri* luminescence test and also showed an increase in toxicity within the first 30 minutes of ozonation followed by an increase over the next 30 minutes of treatment, thus a degradation dependent toxicity was observed [291]. Reported EC_{50} values range from 1.57 mg/L to 210 mg/L depending on the organism used for testing [17, 232, 289, 292]. Differences in toxicity can be due to the use of much higher initial conentrations of Sulfamethoxazole before treatment than used in this study. All these results indicate that different oxidation by-products might be formed depending on the oxidation method used. Additionally a great variance in organism susceptibility might be the reason for these differing effects.

6.9.2. Fluoroquinolone antibiotics

Fluoroquinolone antibiotics are commonly used as broad spectrum-antibiotcs in human as well as veterinary medicine. In human medicine they are an important part during the treatment of nosocomial infections. In addition to their widespread use a general misuse of antibiotics has led not only to the development of multi-resistant bacterial strains but also to their detection in waters of up to 124 µg/L [293]. Fluoroquinolones are of great concern due to their widespread and common use. Two antibiotics of this group (Ciprofloxacin and Ofloxacin) were therefore chosen for comparative analyses regarding their toxicity before and after oxidative treatment.

Discussion

The results of Ciprofloxacin gained in this study (chapter 5.5.2) demonstrate that again the toxicity after oxidation is dependent on the water matrix (e.g. pure water, waste water). While there were no cytotoxic or genotoxic effects after the treatment of WWTP effluent containing Ciprofloxacin, weak cytotoxic effects have been detected after the treatment of HPLC water. Cytotoxic effects of Ciprofloxacin were shown by Gürbay et al. (2005) with Ciprofloxacin concentrations > 50 mg/L in HeLa cells [294]. The *V. fischeri* test performed by Hernando et al. (2007) revealed no toxic effects and the EC_{50} value (> 5.9 mg/L) was above the water solubility [242]. A study with *Giardia lambia* however showed toxic effects of Ciprofloxacin looking at a variety of endpoints [295].

Another endpoint often investigated is the mutagenic activity of Ciprofloxacin. The results presented above (chapter 5.5.2), show that Ciprofloxacin itself as well as the samples after UV/H_2O_2 treatment are not mutagenic in the Ames test using the *S. typhimurium* strains TA98 and TA100. This is in contrast to published data. Clerch et al. (1996) tested the effects of Ciprofloxacin on a variety of *S. typhimurium* strains and found it to be mutagenic [296]. Hartmann et al. (1998) tested hospital waste waters using the *umuC* test and identified Ciprofloxacin as a main source for toxic effects in this test [293].

Like Ciprofloxacin, Ofloxacin is a fluoroquinolone antibiotic. The results of the tests with Ofloxacin (chapter 5.5.3) are similar to those with Ciprofloxacin. No initial toxicity was detected and the tests after ozonation or UV/H_2O_2 treatment did also show no toxic effects in the here used test systems. A literature research by Ferrari et al. (2004) revealed EC_{50} values of acute toxicity tests between 0.01 mg/L and even more than 90 mg/L of Ofloxacin depending on the organism and exposure time used (bacterium, protozoan, algae, crustacean, fish) [297]. Since fluoroquinolones are also commonly used in medical treatment by eye care professionals e.g. for the treatment of bacterial keratitis Bezwada et al. (2008) tested their cytotoxic effects on human corneal keratocytes and endothelial cells. While Ciprofloxacin was cytotoxic at all tested concentrations (10 µg/L – 1 g/L), Ofloxacin only showed statistically significant effects at the highest tested concentration after 15 min of exposure, however an exposure time dependent toxicity (decrease in concentration resulting in effects) was detected [298]. A similar study was performed by Tsai et al. (2010).

Discussion

They also tested the cytotoxic effects of fluoroquinolone antibiotics on human corneal epithelial cells. Toxic effects were also shown to be exposure time dependent, however Ciprofloxacin had stronger cytotoxic effects in all three tests (MTS (3-(4,5-dimethylthiazol-2-yl)-5-(3-carboxymethoxyphenyl)-2-(4-sulfophenyl)-2H-tetrazolium) assay, TEER (Trans Epithelial Electric Resistance) test, cell morphology) than Ofloxacin [299]. In addition the cytotoxicity of Ofloxacin against rabbit corneal epithelial cells was tested by Scuderi et al. (2003). Both test systems, the MTT and the Neutral Red assay, resulted in a dose (1.5 – 6 g/L) and time dependent (8 – 72 h) decrease in cell viability. However the lowest dose (1.5 g/L) already resulted in a statistically significant loss of viability after the shortest exposure time (8 h) [300]. The phototoxic effects of Oflocaxin on HL60 and K 562 cells have been studied by Trisciuoglio et al. (2002) since photosensitizing effects on the skin are a well known side effect during drug therapy. They report photosensitizing effects only in Ofloxacin treated cells (0.05 mM) irradiated with a wavelength of 330 nm [301]. Genotoxic effects of fluoroquinolones were investigated by Itoh et al. (2006) using human lymphoma cells (WTK-1) in the Alkaline Comet Assay. A maximum of 1 g/L of Ofloxacin did not induce DNA damage after up to 20 h of exposure. In contrast the same test with Ciprofloxacin resulted in significant effects which are again exposure time and concentration dependent. The lowest concentration resulting in genotoxic effects was 250 mg/L after 20 hours of exposure. Besides these two, they also tested six other quinolone antibiotics (Naladixic acid, Pipemidic acid, Oxolinic acid, Piromidic acid, Enoxacin and Norfloxacin), however only Norfloxacin also showed an increase in DNA damage after exposure [302]. This study indicated that substances from the same class of antibiotics do not necessarily result in the same cell damaging effects. Since these effects were observed *in vitro*, McQueen et al. (1991) performed a comparative study looking at *in vitro* and *in vivo* genotoxic effects of quinolone antibiotics in rat hepatocytes. Genotoxic effects (unscheduled DNA synthesis (UDS)) were only detected *in vitro* possibly due to a missing metabolism in isolated hepatocytes. However except Naladixic acid, all tested quninolone antibiotics induced UDS, including Ciprofloxacin and Ofloxacin with an effective concentration of 400 µg/mL [303].

Discussion

6.9.3. Triclosan and 2.4-Dichlorophenol

Triclosan is a bactericide and a constituent of disinfection products, preservatives, and everyday products e.g. toothpaste or sports clothing. Although it is found in many different products it raises concern. As well as Bisphenol A, Triclosan has also been added to the CoRAP list because of its persistent, bioaccumulative, and toxic properties [282, 304]. Because of this reason it was selected to be investigated in this study for a better understanding of its behavior during oxidative treatment and the resulting toxicity since it was found in influents of waste water treatment plants with the highest detected concentrations around 80 µg/L [305].

Triclosan has been described as toxic towards a variety of species. The adverse effects of Triclosan include cytotoxic, genotoxic and endocrine properties. These effects have been shown by various groups. Jirasripongpun et al. (2008) tested Triclosan for its cytotoxic and genotoxic effects towards three animal cell lines (BHK-21, Vero and KB cells). Reported IC_{50} values range from 9.8 – 75.28 mg/L depending on the cell line indicating a higher susceptibility to Triclosan of the primate cell lines (Vero and KB) than the rodent cell line (BHK-21). In terms of genotoxicity they report initial effects in the Comet Assay after a three day exposure to 6.66 mg/L which increased after five days of exposure [306]. Genotoxic effects of Triclosan were also described by Binelli et al. (2009) using the *in vitro* Alkaline Comet Assay in zebra mussel hemocytes after an exposure time of 60 minutes. Significant increases in DNA damage were already detected at the lowest tested concentration (28.95 µg/L) [307] which is comparable to the concentrations (0-100 µg/L) used here in this study (chapter 5.5.4), with the first significant effect detected at 1 µg/L after 24 h of exposure. Another Comet Assay was performed by Ciniglia et al. (2005) describing a significant increase in tail moment, thus DNA damage, after an exposure of *Closterium ehrengergii* to 0.25 mg/L Triclosan. Cytotoxic effects were detected at 0.187 mg/L [308]. Zuckerbraun et al. (1998) describe first notable effects in different cytotoxicity tests starting at 2.9 mg/L [309]. As has been shown for other substances EC_{50} values vary between test, exposure time and organisms used. An exposure to Triclosan resulted in EC_{50} values in the range of 0.7 µg/L to 350 µg/L depending on the organism [305, 310-312]. First effects on algae were seen at 0.015 µg/L [313]. Besides its cyto- and genotoxic effects, Triclosan was also reported being an endocrine-disrupting compound in aquatic organisms however at

Discussion

concentrations much higher than commonly found in water bodies [188, 314]. Despite all these reported toxic effects a review on Triclosan and its environmental impacts by Dann et al. (2011) summarizes that the use of Triclosan can be considered as safe, however further research regarding its impact on the ecosystem and the human health needs to be done [171]. A similar statement was published by the Scientific Committee on Consumer Products (2009). They conclude that its use as a preservative is safe, however its use in e.g. body lotions and mouth washs might not be safe because of the high levels of exposure [315]. The US FDA (2008) in addition classified Triclosan as a Category III product due to insufficient information on the safety and effectiveness, especially on missing data of its possible dermal carcinogenicity [316].

2,4-Dichlorophenol is a substance originating not only from the ozonation and photodegradation of Triclosan, but also from its metabolization [173, 317]. In the German surface water guideline, 2,4-Dichlorophenol has been regulated with an environmental qualitiy standard of 10 µg/L. Environmental quality standards are applied when a substances is belived to have a negative impact for the environment or human health [30]. The toxicity of 2,4-Dichlorophenol was amongst others investigated by Chen et al. (2004) using L929 (mouse fibroblasts). The EC_{50} value after a 24 h exposure was 135.29 mg/L. Additionally Chen et al. (2004) were able to show the properties of 2,4-Dichlorophenol to induce DNA fragmentation and apoptosis. Looking at other chlorophenols they also found that the cytotoxicity increased with an increasing number of chlorine atoms [318]. This was also reported by Jiang et al. (2004) who also investigated the cytotoxicity of chlorophenols on L929 but also on HepG2 cells. The LC_{50} values of 127.3 mg/L (L929) and 128.77 mg/L (HepG2) for both cells lines are comparable. In regard to the simultaneously performed QSAR approach correlating the chemical structure and biological activity of 2,4-Dichlorophenol they not only found a relationship of cytotoxicity and the number of chlorine atoms, they also found a connection between toxicity and the position of the chlorine atoms [319]. An increase in chromosomal aberrations after an exposure of CHO cells to 97.8 mg/L (with S9) and 195.6 mg/L (without S9) was shown whereas no statistically significant effects were seen in TK6 cells [320]. The effects of 2,4-Dichlorophenol were also investigated by Ensley et al. (1994). Using the angiosperm *Lemna gibba* for their experiments they detected an EC_{50} of

Discussion

1.5 mg/L and a lethal concentration of 6.52 mg/L [321]. Own tests with 2,4-Dichlorophenol demonstrate that neither cytotoxic nor genotoxic effects have been detected at concentrations between 0 and 100 µg/L (chapter 5.5.4). Comparing these results to the environmental quality standard (EQS) of 10 µg/L it can be said that this value seems to be a safe measure.

Overall it can be concluded that Triclosan can be removed from water by ozonation and that the ozonation leads to the formation of the less genotoxic by-product 2,4-Dichlorophenol. However an EQS for 2,4-Dichlorophenol has been established and its formation should therefore be avoided.

6.10. Biocides

6.10.1. Irgarol 1051

Irgarol® 1051 is a constituent of paints used e.g. on ship hulls to prevent algal growth or on facades reaching waste water treatment plants through leaching and wahs-off into the sewage system [322]. Therefore its presence in the water system is likely (up to 2.5 µg/L in freshwater [323]) and a reason it was investigated in this project.

Own results showed that although the ozonation of Irgarol led to the formation of different oxidation by-products [151] no cytotoxic or genotoxic effects were detected before or after ozonation. In addition no differences in cell viability or Olive Tail Moment were observed comparing the use of 56 µg or 400 µg ozone (chapter 5.6.1).

Published studies showed that the toxicity of Irgarol 1051 has been investigated using a variety of aquatic organisms, since they are most affected by its presence in water systems. Although Irgarol is produced as a biocide effects in non-target organisms should be avoided. EC_{50} values gained from the *V. fischeri* luminescence test range from 2.42 mg/L to more than 50 mg/L depending on the exposure time [227, 228, 324-326]. Irgarol has also been tested for its effects on *D. magna* resulting in EC_{50} values of 7.3 or 8.3 mg/L [227, 228, 324, 325] and in the microalgae *Selenastrum capricronotum* with EC_{50} values between 0.01 and 15.5 mg/L [174, 228,

Discussion

324, 327]. Mesocosm studies by Mohr et al. (2009) revealed direct effects on the used organisms after only one single application of Irgarol. The most sensitive organism from this mesocosm was *Myriophyllum verticillatum* with an EC_{50} of 0.21 µg/L after 150 days of exposure [323]. A study performed by Noguerol et al. (2006) investigated the interactions of vertebrate receptors (estrogen and aryl hydrocarbon receptors) with a variety of pollutants including Irgarol 1051. Their results show that Irgarol 1051 did not lead to a response (inhibition or activation) in those yeast based test systems. Since they did also not detect any effects for the structurally similar Sea-Nine biocide they concluded that heterocyclic substances are not able to interact with estrogen or aryl hydrocarbon receptors [328].

Other studies also focused on M1 (2-Methylthio-4-*tert*-butylamino-6-amino-*s*-triazine) a degradation product formed through biological or chemical degradation as well as through photodegradation of Irgarol® 1051 [329-331]. Okamura et al. (2003) looked at a variety of organisms, to detect differences in their susceptibility to Irgarol and its degradation product M1 and found differences in EC_{50} values [327] which was also reported by Fernandez-Alba et al. [228]. The results of both studies are displayed in Table 32. Their results also show great differences in organism susceptibility towards Irgarol® 1051 or M1 with a huge range of EC_{50} values (0.0081 mg/L - > 50 mg/L for Irgarol® 1051 and 0.0071 mg/L - > 50 mg/L for M1). It can also be seen, that for some organisms (e.g. *L. sativa* or *V. fischeri*) the EC_{50} of M1 is lower than the one for Irgarol® 1051 thus indicating an increase in toxicity due to the formation of a more toxic transformation product.

Discussion

Table 32: Summary of EC_{50} or LC_{50} values of Irgarol 1051 and M1 found in the literature

Author	Organism (Exposure time)	Irgarol® 1051 EC_{50} [mg/L]	M1 EC_{50} [mg/L]	Reference
Okamura et al.	S. capricornotum (3 d)	1.6	19	[325, 327]
	L. gibba (7 d)	0.11	0.12	
	L. minor (7 d)	0.0081	0.0071	
	L. sativa (5 d)	> 50	4.3	
	V. fischeri (30 min)	> 50	> 50	
	D. magna (48 h)	8.3 (LC_{50})	11 (LC_{50})	
	D. duplex (24 h)	5.7 (LC_{50})	27 (LC_{50})	
	T. platyurus (24 h)	12 (LC_{50})	19 (LC_{50})	
	A. salina (24 h)	> 40	> 40	
Fernandez-Alba et al.	V. fischeri (15 min)	50.8	6.5	[228]
	S. capricornotum (30 h)	15.5	4.7	
	D. magna (48 h)	7.3	18.9	

6.10.2. Terbutryn

Terbutryn is a biocide and as Irgarol® 1051 it is used in paints for facades or ship hulls. It is also used as a control agent for grasses and weeds which might also lead to a washing-off into surface waters. Due to rain events Terbutryn might leach directly into surface water or the sewer system and after an incomplete removal in a waste water treatment plant it might also reach surface waters [322]. Velisek et al. (2010) reported a concentration of 5.6 µg/L detected in surface waters.

For Terbutryn the above presented results of this study show an influence of the ozone concentration on the toxicity (chapter 5.6.2). It could be shown that the substance itself already has genotoxic effects. Depending on the O_3 concentration the effects became even stronger indicated by an increase in the Olive Tail Moment compared to the untreated samples but genotoxic effects were completely removed at the highest O_3 concentration (800 µg). This indicates that using 800 µg O_3 the genotoxic Terbutryn had been degraded without the formation of toxic oxidation by-

Discussion

products which have only been formed at the lower ozone concentrations (36 – 195 µg). However ozone concentrations commonly applied during waste water treatment are usually higher than 800 µg. Thus there probably is no risk in applying ozone to waters containing Terbutryn since the toxic Terbutryn is removed by ozone and the formed genotoxic oxidation by-products only occur at low O_3 concentrations.

The toxicity of Terbutryn itself has been shown by others before, however with differing results. Arufe et al. (2004) observed a LC_{50} value of 1.4 mg/L using *Sparus aurata* larvae (gilt-head bream) while the EC_{50} value derived from *Vibrio fischeri* was 15.94 mg/L [332]. In contrast, Hernando et al. (2007) classified Terbutryn as non-harmful for *Vibrio fischeri* according to the toxicity categories of the EU legislation [242]. The general toxicity of Terbutryn was also tested using *V. fisheri* by Gaggi et al. (1995) resulting in an EC_{50} of 13 mg/L, comparable to the published data by Arufe et al. In comparison, the EC_{50} values for two algae were 2.7 mg/L (*Selenastrum capricornutum*) or 3.1 mg/L (*Dunaliella tertiolecta*) and 22 mg/L for the brine shrimp *Artemia salina* [333]. Genotoxic effects of Terbutryn in human peripheral blood leukocytes were detected at a concentration of 5 mg/L using the Alkaline Comet Assay [176, 334] whereas no effects were seen in the Sister Chromatid Exchange and in the Micronucleus Assay testing concentrations up to 150 mg/L [176]. Toxic effects of Terbutryn on aquatic organisms were tested by Velisek et al. (2010) and Plhalova et al. (2010) LC_{50} values for juvenile *Poecilia reticulata* and *Danio rerio* (*D. rerio*) were 2.85 mg/L and 5.71 mg/L whereas the LC_{50} value for embryonic *D. rerio* was determined as 8.04 mg/L, thus showing great differences in susceptibility according to age and organism [335]. Velisek at al. investigated various parameters (hematology, histology, blood, biochemistry and biometric parameters) in carp exposed to Terbutryn and found effects starting at 0.04 mg/L [336] and even lower effective concentrations (0.02 µg/L) on the early life stages of carp [337].

The above described results (chapter 5.6.2) from this PhD study clearly show an increase in genotoxic effects after ozonation of Terbutryn which can be related to formed oxidation by-products. Brix et al. (2009) also found an increase in toxicity after oxidative treatment of Terbutryn. In their study they looked at disinfection by-products after hypochlorite (HClO) treatment. The toxicity was determined by the *Vibrio fischeri* luminescence test and showed a decrease of the EC_{50} value of

Discussion

Terbutryn from 24 mg/L in untreated water to 15 mg/L in HClO treated water thus they concluded the formation of disinfection by-products which are more toxic than Terbutryn itself [338].

Own results as well as results from other studies demonstrated the formation of toxic by-products resulting from the oxidation of Terbutryn. Terbutryn has toxic effects even at low concentrations and its presence in water bodies therefore is a risk for the environment. However higher amounts of oxidant seem to be effective in removing Terbutryn as well as its toxic by-products and lower its risk potential.

6.11. Musk fragrances

6.11.1. AHTN and HHCB

Musk fragrances like AHTN and HHCB are aromatic substances ubiquitously used in cosmetics, perfumery and detergents. Since they have a low water solubility they easily bioaccumulate. This characteristic has led to their detection in the environment (up to 45 µg/L [1]) as well as human blood samples. The use of musk fragrances in cosmetics etc. has been classified as safe however their behavior and fate in the environment needs further investigation [339].

Comparing own results of AHTN (chapter 5.7.1) before, during, and after ozonation or UV/H_2O_2 treatment it can be seen, that the ozonation does not result in the formation of toxic by-products, whereas after 20 min of UV/H_2O_2 treatment weak cytotoxic effects occur. These effects are related to formed oxidation by-products, since no remaining peroxides were detected. Although oxidation by-products were also formed during ozonation [186] no cytotoxic or genotoxic effects were observed. HHCB-lactone was tested for toxicity at concentrations between 0 µg/L and 50 µg/L but neither cytotoxic nor genotoxic effects occurred. Cytotoxic effects however, were only detected after 60 min UV/H_2O_2 oxidation of 0.1 mg/L HHCB. This reflects the findings for Bisphenol A (chapter 5.4.2), where only the UV/H_2O_2 treatment resulted in a reduced viability, indicating that different oxidation by-products are formed depending on the applied oxidation method. The toxicity of both substances, HHCB

Discussion

and its oxidation by-product HHCB-lactone, has been studied before by others and resulted in a classification as not harmful for the environment based on PNEC (predicted non effect concentration) values for soil dry weight and for aquatic species [340]. The genotoxicity tests performed in this study confirm previously published results by Api et al. (1999) who did not find genotoxic or mutagenic effects of AHTN and HHCB in different test systems (Ames Test, Chromosome abberations, Unscheduled DNA synthesis and Micronucleus test) [341]. Effects of AHTN and HHCB on the early life stages of zebrafish have been investigated by Carlsson et al. (2004). While AHTN had effects on the heart rate (LOEC = 33 µg/L) no effects were seen on the survival time with a maximum concentration of 100 µg/L. HHCB in contrast had no effects in both tests up to a concentration of 1 mg/L [342]. However, Randelli et al. (2011) reported effects of AHTN on the regulation of immunoregulatory genes in gonadal cells of the rainbrow trout, while there were no effects on cell viability [343].

In addition to the cytotoxicity and genotoxicity testing in this study HHCB and HHCB-lactone were also tested for estrogenicity (chapter 5.7.1 and 5.7.2), but no effects were detected supporting other published results. Estrogenic effects of AHTN and HHCB have been studied by Seinen et al. (1999) They found effects on the uterus weight of mice starting at 50 ppm AHTN and an induction of the ERα transcription starting at 50 µM AHTN. However, they concluded that compared to effects resulting from a 17ß-estradiol exposure the effects induced by AHTN are not noteworthy. Thus, they also state that AHTN does not have a negative effect on the environment [344]. Schreurs et al. (2005) investigated the interactions of musk fragrances with different hormonal receptors and found AHTN as well as HHCB to be antagonists towards the ERß receptor [345] which confirms the above presented findings of our own investigations where no estrogenic activity of HHCB was detected. Antiestrogenic effects of HHCB were also reported by Simmons et al. (2012) after performing the Yes-assay, as well as other *in* vitro and *in* vivo test systems [346]. A set of different reporter gene assays (estrogen, androgen, progesterone, and glucocorticoid) was used to test HHCB and AHTN for their potential endocrine effects. Here again it was reported that neither substance had agonistic effects whereas antagonistic effects were detected in each assay except the glucocorticoid system [347]. Another study by Schreurs et al. (2003) also proves the antiestrogenic

Discussion

effect, this time using a transgenic zebrafish assay [348]. The same findings on the antagonistic mode of action by polycyclic musks have been reported in a literature review by Witorsch and Thomas (2010) [349]. Although HHCB has been shown to act estrogenic, the concentrations inducing these effects were much higher (2.58 mg/L) than those used in our study (max. 0.1 mg/L) [350]. Schreurs et al. (2002 and 2005) were also able to prove, that the endocrine mode of action (agonistic or antagonistic) of HHCB is cell line dependent, and HHCB does not always result in estrogenic effects [345, 351]. Despite the published data on the toxicity and endocrine activity of HHCB and AHTN further analyses especially in regard to the formation of oxidation by-products should be performed since both substances are frequently detected in water systems [36, 352].

6.12. Organophosphates: TCEP, TCPP and TPP

Organophosphates are substances commonly used as constituents of biocides and thus are designed to be biologically active. Besides this they are also used as flame retardants and might therefore reach water systems. Although despite their widespread use, toxicological data about the three here used organophosphates is scarce.

TCEP has been added to the candidate list of substances of very high concern (SVHC) by the ECHA due to its classification as a reproductive toxicant [180]. The SVHC supporting document also indicates neurotoxic effects after an exposure to TCEP [353].

The toxicity of TPP has been investigated by Lin et al. (2009). They report LC_{50} values of 0.51 mg/L and 0.09 mg/L for *Daphnia magna* exposed to TPP for 24 h or 48 h [221]. Föllmann et al. (2006) tested TCEP and TCPP for their potential to induce cytotoxic, genotoxic, mutagenic, and estrogenic effects *in vitro*. The results show, that an exposure of V79 cells (hamster fibroblasts) to both substances did not result in genotoxic and mutagenic effects using the Alkaline Comet Assay and the

Discussion

Ames test with and without metabolic activation (S9). In addition no estrogenic or anti-estrogenic effects were detected for a maximum concentration of 1 mM (TCEP = 250 mg/L; TCPP = 327.55 mg/L). However, cytotoxic effects for both substances were seen at concentrations above 10 µM for TCEP and 1 mM for TCPP but only in the presence of the S9 liver homogenate [354]. Similar results for genotoxic or mutagenic effects were also published by others [355-358]. Ren et al. (2009) were able to show cytotoxic effects to start at 1 mg/L in the LDH assay using primary rat renal cells. At these concentrations TCEP also had an influence on the ion uptake and the expression of ion-transporters [359]. Another study of the same group also showed possible risks on the ecosystem and the human health [360]. Endocrine effects of organophsophates on H295R and MVLN cells have been investigated by Liu et al. (2012) demonstrating the potential of endocrine disruption of TCEP, TCPP and TPP as well as other organophosphates starting at low µg/L concentrations. At the same time they were able to show differences in the mode of action [361].

Comparing the published data for TCEP and TCPP to results obtained during this study it can be seen, that effects occur at mainly much higher concentrations than tested here (chapter 5.8.2 and 5.8.3). 0.1 mg/L of the organophosphates did not result in cytotoxic or genotoxic effects. Cytotoxic effects were only seen after UV/H_2O_2 treatment but not after ozonation, indicating the formation of toxic oxidation by-products.

In contrast to the above mentioned data from the literature, Higley et al. (2012) performed a study with sediments from the upper Danube river using an effect-directed analysis looking at endocrine disrupting, mutagenic, and teratogenic effects. Although they found toxic effects for some fractions of the extracted sediments, the fractions containing organophosphates did not result in any effects [362].

The results for TPP however are different. 0.1 mg/L TPP did not result in a decrease in cell viability neither before nor after ozonation. Genotoxic effects however were detected before oxidative treatment but not after 60 min of treatment. Thus ozonation and UV/H_2O_2 oxidation were useful to remove the toxic TPP from water samples. Flaskos et al. (1994) report an IC_{50} value of 250 mg/L for TPP in PC12 cells (rat pheochromocytoma cells) [363]. This IC_{50} value in fact is a lot higher than the concentration tested here.

Discussion

In conclusion it can be said that these results show that substances belonging to the same class of chemicals do not necessarily exhibit the same mode of action and oxidative treatment might result in different outcomes.

Discussion

6.13. Conclusions and future directions

The results of this study also demonstrate that a generalized statement about toxic effects of waste water treatment plant effluents is not possible. The water composition as well as the waste water treatment plant design and the operating conditions might vary between different plants and over time. In addition four *in vitro* tests have been applied in this study and most of the substances and treatment conditions resulted in no toxic effects. However, other toxicological test methods as well as *in vivo* methods might give different results, due to a different mode of action or in case of *in vivo* methods, the metabolization of micropollutants. Therefore further research (e.g. EDA) regarding the toxicity of micropollutants, their metabolites as well as their oxidation by-products is needed especially since up to now the methods of waste water treatment have not been applied or modified in regard to micropollutant removal.

Another step towards a better surveillance of water quality by means of toxicity is the development or establishment of existing online monitoring systems, for a faster on-site analysis in combination with other test systems.

Further technologies that might also be applied for a risk assessment of organic micropollutants in water systems are the QSAR-approaches or toxicogenomics as well as biosensors which allow a toxicological analysis at the source, e.g. directly at the WWTP effluent discharge point or even in front of it. The use of the so called "–omics" techniques might be another step towards the application of cellular based test systems. These techniques allow an analysis on the whole organism situation e.g. changes in DNA expression (genomics) or protein profiles (proteomics) [231]. For further or more detailed analyses *in vivo* methods might then be applied after limiting the vast amount of adverse effects through *in vitro* testing. Concepts like the Adverse Outcome Pathways proposed by the US EPA (2010), effect directed analysis or procedures as suggested by Kase et al. (2009) are most promising for a secure risk assessment [71, 234, 236].

.

7. References

1. Carballa, M., F. Omil, J.M. Lema, M. Llompart, C. Garcia-Jares, I. Rodriguez, M. Gomez, and T. Ternes, *Behavior of pharmaceuticals, cosmetics and hormones in a sewage treatment plant.* Water Research, 2004. **38**(12): p. 2918-2926.
2. Fick, J., H. Söderström, R.H. Lindberg, C. Phan, M. Tysklind, and D.G.J. Larsson, *Contamination of surface, ground, and drinking water from pharmaceutical production.* Environmental Toxicology and Chemistry, 2009. **28**(12): p. 2522-2527.
3. Hirsch, R., T. Ternes, K. Haberer, and K.L. Kratz, *Occurrence of antibiotics in the aquatic environment.* Science of the Total Environment, 1999. **225**(1-2): p. 109-118.
4. Tambosi, J.L., L.Y. Yamanaka, H.J. Jose, R. Moreira, and H.F. Schroder, *Recent Research Data on the Removal of Pharmaceuticals from Sewage Treatment Plants (STP).* Quimica Nova, 2010. **33**(2): p. 411-420.
5. Vieno, N., T. Tuhkanen, and L. Kronberg, *Elimination of pharmaceuticals in sewage treatment plants in Finland.* Water Research, 2007. **41**(5): p. 1001-1012.
6. Henze, M., Harremoes, P., la Cour Jansen, J., Arvin, E., *Wastewater Treatment: Biological and Chemical Processes.* Springer, 2002. **3. Edition**.
7. Tauxe-Wuersch, A., L.F. De Alencastro, D. Grandjean, and J. Tarradellas, *Occurrence of several acidic drugs in sewage treatment plants in Switzerland and risk assessment.* Water Research, 2005. **39**(9): p. 1761-1772.
8. Ternes, T.A., J. Stuber, N. Herrmann, D. McDowell, A. Ried, M. Kampmann, and B. Teiser, *Ozonation: a tool for removal of pharmaceuticals, contrast media and musk fragrances from wastewater?* Water Research, 2003. **37**(8): p. 1976-1982.
9. Nikolaou, A., S. Meric, and D. Fatta, *Occurrence patterns of pharmaceuticals in water and wastewater environments.* Analytical and Bioanalytical Chemistry, 2007. **387**(4): p. 1225-1234.
10. Rosal, R., A. Rodriguez, J. Antonio Perdigon-Melon, A. Petre, E. Garcia-Calvo, M. Jose Gomez, A. Aguera, and A.R. Fernandez-Alba, *Occurrence of emerging pollutants in urban wastewater and their removal through biological treatment followed by ozonation.* Water Research, 2010. **44**(2): p. 578-588.
11. Schaar, H., Kreuzinger, N., *Endbericht KomOzon: Technische Umsetzung und Implementierung einer Ozonungsstufe für nach dem Stand der Technik gereinigtes kommunales Abwasser Heranführung an den Stand der Technik.* Lebensministerium Wien, 2011.
12. Bergmann, A., Fohrmann, R., Weber, F./A., *Zusammenstellung von Monitoringdaten zu Umweltkonzentrationen von Arzneimitteln.* Umweltbundesamt, 2011.
13. Ternes, T., *Occurrence of drugs in German sewage treatment plants and rivers.* Water Research, 1998. **32**(11): p. 3245-3260.
14. Abegglen, C., *Ozonung von gereinigtem Abwasser Schlussbericht Pilotversuch Regensdorf.* 2009.
15. Rizzo, L., S. Meric, D. Kassinos, M. Guida, F. Russo, and V. Belgiorno, *Degradation of diclofenac by TiO2 photocatalysis: UV absorbance kinetics*

References

and process evaluation through a set of toxicity bioassays. Water Research, 2009. **43**(4): p. 979-988.

16. Sein, M.M., T.C. Schmidt, A. Golloch, and C. von Sonntag, *Oxidation of some typical wastewater contaminants (tributyltin, clarithromycin, metoprolol and diclofenac) by ozone.* Water Science and Technology, 2009. **59**(8): p. 1479-1485.
17. Wammer, K.H., T.M. Lapara, K. McNeill, W.A. Arnold, and D.L. Swackhamer, *Changes in antibacterial activity of triclosan and sulfa drugs due to photochemical transformations.* Environmental Toxicology and Chemistry, 2006. **25**(6): p. 1480-1486.
18. Hollender, J., S.G. Zimmermann, S. Koepke, M. Krauss, C.S. McArdell, C. Ort, H. Singer, U. von Gunten, and H. Siegrist, *Elimination of Organic Micropollutants in a Municipal Wastewater Treatment Plant Upgraded with a Full-Scale Post-Ozonation Followed by Sand Filtration.* Environmental Science & Technology, 2009. **43**(20): p. 7862-7869.
19. Reungoat, J., M. Macova, B.I. Escher, S. Carswell, J.F. Mueller, and J. Keller, *Removal of micropollutants and reduction of biological activity in a full scale reclamation plant using ozonation and activated carbon filtration.* Water Research, 2010. **44**(2): p. 625-637.
20. Zwiener, C. and F.H. Frimmel, *Oxidative treatment of pharmaceuticals in water.* Water Research, 2000. **34**(6): p. 1881-1885.
21. Bauer, R. and H. Fallmann, *The Photo-Fenton Oxidation — A cheap and efficient wastewater treatment method.* Research on Chemical Intermediates, 1997. **23**(4): p. 341-354.
22. Macova, M., B.I. Escher, J. Reungoat, S. Carswell, K.L. Chue, J. Keller, and J.F. Mueller, *Monitoring the biological activity of micropollutants during advanced wastewater treatment with ozonation and activated carbon filtration.* Water Research, 2010. **44**(2, Sp. Iss. SI): p. 477-492.
23. Vieno, N.M., H. Harkki, T. Tuhkanen, and L. Kronberg, *Occurrence of pharmaceuticals in river water and their elimination in a pilot-scale drinking water treatment plant.* Environmental Science & Technology, 2007. **41**(14): p. 5077-5084.
24. Stalter, D., A. Magdeburg, and J. Oehlmann, *Comparative toxicity assessment of ozone and activated carbon treated sewage effluents using an in vivo test battery.* Water Research, 2010. **44**(8): p. 2610-2620.
25. Petala, M., P. Samaras, A. Zouboulis, A. Kungolos, and G.P. Sakellaropoulos, *Influence of ozonation on the in vitro mutagenic and toxic potential of secondary effluents.* Water Research, 2008. **42**(20): p. 4929-4940.
26. Escher, B., Leusch F., *Bioanalytical tools for safe water supplies: using cellular response to show toxicity effect.* Water21, 2011: p. 31-33.
27. Kavlock, R.J., C.P. Austin, and R.R. Tice, *Toxicity Testing in the 21st Century: Implications for Human Health Risk Assessment.* Risk Analysis, 2009. **29**(4): p. 485-487.
28. *EU WRRL (2000): Richtlinie 2000/60/EG des Europäischen Parlaments und des Rates vom 23. Oktober 2000 zur Schaffung eines Ordnungsrahmens für Maßnahmen der Gemeinschaft im Bereich der Wasserpolitik.*
29. *Richtlinie 2008/105/EG des Europäischen Parlaments und des Rates vom 16. Dezember 2008 über Umweltqualitätsnormen im Bereich der Wasserpolitik.*

References

30. Verordnung zum Schutz der Oberflächengewässer (Oberflächengewässerverordnung - OGewV); Ausfertigungsdatum: 20.07.2011.
31. Grohmann, A.N., Jekel, M., Grohmann, A., Szewzyk, R., Szewzyk, U., *Wasser: Chemie, Mikrobiologie und nachhaltige Nutzun*. de Gruyter, 2011. **ISBN: 978-3-11-021308-9**.
32. Schwarzenbach, R.P., B.I. Escher, K. Fenner, T.B. Hofstetter, C.A. Johnson, U. von Gunten, and B. Wehrli, *The challenge of micropollutants in aquatic systems*. Science, 2006. **313**(5790): p. 1072-1077.
33. *Rote Liste® 2011: Arzneimittelverzeichnis für Deutschland (einschließlich EU-Zulassungen und bestimmter Medizinprodukte)*. Rote Liste Service GmbH, 2011. **ISBN: 978-3-939192-50-3**.
34. Schwabe, U., Paffrath, D., *Arzneiverordnungs-Report 2011; Aktuelle Daten, Kosten, Trends und Kommentare*. Springer, 2011. **ISBN: 978-3-642-21991-7**.
35. WHO, *WHO Model List of Essential Medicines*. 17th list, 2011.
36. Bester, K., *Polycyclic musks in the Ruhr catchment area - transport, discharges of waste water, and transformations of HHCB, AHTN and HHCB-lactone*. Journal of Environmental Monitoring, 2005. **7**(1): p. 43-51.
37. Chen, X. and K. Bester, *Determination of organic micro-pollutants such as personal care products, plasticizers and flame retardants in sludge*. Analytical and Bioanalytical Chemistry, 2009. **395**(6): p. 1877-1884.
38. Heberer, T., S. Gramer, and H.J. Stan, *Occurrence and distribution of organic contaminants in the aquatic system in Berlin. Part III: Determination of synthetic musks in Berlin surface water applying solid-phase microextraction (SPME) and gas chromatography-mass spectrometry (GC-MS)*. Acta Hydrochimica Et Hydrobiologica, 1999. **27**(3): p. 150-156.
39. Rüffer, H., Masannek, R., *Wasser: Nutzung im Kreislauf Hygiene, Analyse und Bewertung (Höll, K., Grohmen, A.)*. de Gruyter, 2002. **8. Edition**.
40. *Abwasserverordnung in der Fassung der Bekanntmachung vom 17. Juni 2004 (BGBl. I S. 1108, 2625), die zuletzt durch Artikel 20 des Gesetztes vom 31. Juli 2009 (BGBl. I S. 2585) geändert worden ist*. 2009.
41. Cooper, P.F., *"Historical aspects of wastewater treatment" in Decentralised Sanitation and Reuse: Concenpts, systems and implementation*. IWA Publishing, 2001.
42. Judd, S., Judd, C., *The MBR Book: Principles and Applications of Membrane Bioreactors for Water and Wastewater Treatment*. EISEVIER Butterworth-Heinemann, 2011. **2. Edition**: p. ISBN: 978-0-08-096682-3.
43. Bartel, H., *Wasser: Nutzung im Kreislauf Hygiene, Analyse und Bewertung (Höll, K., Grohmen, A.)*, de Gruyter, 2002. **8. Edition**.
44. Zimmermann, S.G., M. Wittenwiler, J. Hollender, M. Krauss, C. Ort, H. Siegrist, and U. von Gunten, *Kinetic assessment and modeling of an ozonation step for full-scale municipal wastewater treatment: Micropollutant oxidation, by-product formation and disinfection*. Water Research, 2011. **45**(2): p. 605-617.
45. Cecen, F., Aktas, Ö., *Activated Carbon for Water and Wastewater Treatment: Integration of Adsorption and Biological Treatment*. Wiley-VCH, 2011. **ISBN: 978-3-527-32471-2**.
46. Anderson, D., *Activated Carbon Treatment of Waste Waters*. Effluent & Water Treatment Journal, 1971. **11**(3): p. 144-&.

References

47. Serrano, D., S. Suárez, J.M. Lema, and F. Omil, *Removal of persistent pharmaceutical micropollutants from sewage by addition of PAC in a sequential membrane bioreactor.* Water Research, 2011. **45**(16): p. 5323-5333.
48. Serrano, D., J.M. Lema, and F. Omil, *Influence of the employment of adsorption and coprecipitation agents for the removal of PPCPs in conventional activated sludge (CAS) systems.* Water Science and Technology, 2010. **62**(3): p. 728-735.
49. Tomaszewska, M. and S. Mozia, *Removal of organic matter from water by PAC/UF system.* Water Research, 2002. **36**(16): p. 4137-4143.
50. Westerhoff, P., Y. Yoon, S. Snyder, and E. Wert, *Fate of Endocrine-Disruptor, Pharmaceutical, and Personal Care Product Chemicals during Simulated Drinking Water Treatment Processes.* Environmental Science & Technology, 2005. **39**(17): p. 6649-6663.
51. Andreozzi, R., V. Caprio, A. Insola, and R. Marotta, *Advanced oxidation processes (AOP) for water purification and recovery.* Catalysis Today, 1999. **53**(1): p. 51-59.
52. Ikehata, K., M. Gamal El-Din, and S.A. Snyder, *Ozonation and advanced oxidation treatment of emerging organic pollutants in water and wastewater.* Ozone-Science & Engineering, 2008. **30**(1): p. 21-26.
53. Tuerk, J., B. Sayder, A. Boergers, H. Vitz, T.K. Kiffmeyer, and S. Kabasci, *Efficiency, costs and benefits of AOPs for removal of pharmaceuticals from the water cycle.* Water Science and Technology, 2010. **61**(4): p. 985-993.
54. Somensi, C.A., E.L. Simionatto, S.L. Bertoli, A. Wisniewski, Jr., and C.M. Radetski, *Use of ozone in a pilot-scale plant for textile wastewater pre-treatment: Physico-chemical efficiency, degradation by-products identification and environmental toxicity of treated wastewater.* Journal of Hazardous Materials, 2010. **175**(1-3): p. 235-240.
55. Huber, M.M., A. Gobel, A. Joss, N. Hermann, D. Loffler, C.S. McArdell, A. Ried, H. Siegrist, T.A. Ternes, and U. von Gunten, *Oxidation of pharmaceuticals during ozonation of municipal wastewater effluents: A pilot study.* Environmental Science & Technology, 2005. **39**(11): p. 4290-4299.
56. Kurokawa, Y., A. Maekawa, M. Takahashi, and Y. Hayashi, *Toxicity and Carcinogenicity of Potassium Bromate - A New Renal Carcinogen.* Environmental Health Perspectives, 1990. **87**: p. 309-335.
57. Siddiqui, M.S., G.L. Amy, and R.G. Rice, *Bromate Ion Formation - A Critical-Review.* Journal American Water Works Association, 1995. **87**(10): p. 58-70.
58. Glaze, W.H., J.W. Kang, and D.H. Chapin, *The Chemistry of Water-Treatment Processes Involving Ozone, Hydrogen-Peroxide and Ultraviolet-Radiation.* Ozone-Science & Engineering, 1987. **9**(4): p. 335-352.
59. Hoigné, J. and H. Bader, *Rate constants of reactions of ozone with organic and inorganic compounds in water—II: Dissociating organic compounds.* Water Research, 1983. **17**(2): p. 185-194.
60. Parsons, S.A., Williams, M., *Advanced Oxidation Processes for Water and Wastewater Treatment.* IWA Publishing, 2004. **1.** Edition(ISBN: 14 84339 017 5).
61. Gottschalk, C., Libra, .A., Saupe, A., *Ozonation of Water and Waste Water: A Practical Guide to Understanding Ozone and its Applications.* Wiley-VCH, 2010. **2.** Edition(ISBN: 978-3-527-31962-6).

References

62. DIN 5031-7 (1984-01-00): *Strahlungsphysik im optischen Bereich und Lichttechnik - Teil 7: Benennung der Wellenlängenbereiche.*
63. Barbusinski, K., *Fenton Reaction - Controversy Concerning The Chemistry.* Ecological Chemistry and Engineering S-Chemia I Inzynieria Ekologiczna S, 2009. **16**(3): p. 347-358.
64. Bauer, R., *Applicability of Solar Irradiation for Photochemical Waste-Water Treatment.* Chemosphere, 1994. **29**(6): p. 1225-1233.
65. Ruppert, G., R. Bauer, and G. Heisler, *The Photo-Fenton Reaction - An Effective Photochemical Waste-Water Treatment Process.* Journal of Photochemistry and Photobiology a-Chemistry, 1993. **73**(1): p. 75-78.
66. Trovo, A.G., R.F.P. Nogueira, A. Aguera, A.R. Fernandez-Alba, C. Sirtori, and S. Malato, *Degradation of sulfamethoxazole in water by solar photo-Fenton. Chemical and toxicological evaluation.* Water Research, 2009. **43**(16): p. 3922-3931.
67. Ruppert, G., R. Bauer, and G. Heisler, *UV-O_3, UV-H_2O_2, UV-TIO_2 and the Photo-Fenton Reaction - Comparison of Advanced Oxidation Processes for Waste-Water Treatment.* Chemosphere, 1994. **28**(8): p. 1447-1454.
68. Li, W., V. Nanaboina, Q. Zhou, and G.V. Korshin, *Effects of Fenton treatment on the properties of effluent organic matter and their relationships with the degradation of pharmaceuticals and personal care products.* Water Research, 2012. **46**(2): p. 403-412.
69. De la Cruz, N., J. Gimenez, S. Esplugas, D. Grandjean, L.F. de Alencastro, and C. Pulgarin, *Degradation of 32 emergent contaminants by UV and neutral photo-fenton in domestic wastewater effluent previously treated by activated sludge.* Water Research, 2012. **46**(6): p. 1947-57.
70. Oppenländer, T., *Photochemical Purification of Water and Air.* Wiley-VCH, 2002. **1. Edition**(ISBN: 978-3-527-30563-6).
71. Kase, R., Kunz, P., Gerhardt, A, *Identifikation geeigneter Nachweismöglichkeiten von hormonaktiven und reproduktionstoxischen Wirkungen in aquatischen Ökosystemen.* Umweltwissenschaften und Schadstoff-Forschung, 2009. **21**(4): p. 339-378.
72. Grummt, T., *Wasser: Nutzung im Kreislauf Hygiene, Analyse und Bewertung (Höll, K., Grohmen, A.).* de Gruyter, 2002. **8. Edition**.
73. Umweltbundesamt, *Bewertung der Anwesenheit teil- oder nicht bewertbarer Stoffe im Trinkwasser aus gesundheitlicher Sicht: Empfehlung des Umweltbundesamtes nach Anhörung der Trinkwasserkommission beim Umweltbundesamt.* Bundesgesundheitsbl - Gesundheitsforsch - Gesundheitsschutz, 2003. **46**: p. 249-251.
74. Krewski, D., D. Acosta, M. Andersen, H. Anderson, J.C. Bailar, K. Boekelheide, R. Brent, G. Charnley, V.G. Cheung, S. Green, K.T. Kelsey, N.I. Kerkvliet, A.A. Li, L. McCray, O. Meyer, R.D. Patterson, W. Pennie, R.A. Scala, G.M. Solomon, M. Stephens, J. Yager, L. Zeise, and A. Staff Comm Toxicity Testing, *Toxicity Testing in the 21ST Century: A Vision and a Strategy.* Journal of Toxicology and Environmental Health-Part B-Critical Reviews, 2010. **13**(2-4): p. 51-138.
75. *The European Parliament and the Council of the European Union (2006); REGULATION (EC) No 1907/2006 OF THE EUROPEAN PARLIAMENT AND OF THE COUNCIL of 18 December 2006 concerning the Registration, Evaluation, Authorisation and Restriction of Chemicals (REACH), establishing a European Chemicals Agency, amending Directive 1999/45/EC and*

References

repealing Council Regulation (EEC) No 793/93 and Commission Regulation (EC) No 1488/94 as well as Council Directive 76/769/EEC and Commission Directives 91/155/EEC, 93/67/EEC, 93/105/EC and 2000/21/EC.

76. http://ecvam.jrc.it/index.htm.
77. EU, Directive 2012/63/EU OF THE EUROPEAN PARLIAMENT AND OF THE COUNCIL of 22 September 2010 on the protection of animals used for scientific purposes. 2010.
78. 3R Research Foundation Switzerland: Annual Report 2010. 2010.
79. http://www.bfr.bund.de/de/zebet-1433.html.
80. NIH, Guidance Document on using In Vitro Data to Estimate In Vivo Starting Doses for Acute Toxicity. NIH Publication No: 01-4500, 2001.
81. Barile, F.A., P.J. Dierickx, and U. Kristen, In-vitro Cytotoxicity Testing for Prediction of Acute Human Toxicity. Cell Biology and Toxicology, 1994. **10**(3): p. 155-162.
82. Greim, H., Dreml, E., Toxikologie. Eine Einführung für Naturwissenschaftler und Mediziner. Wiley-VCH, 1996.
83. Eisenbrand, G., Netzler, M., Hennecke, F.J., Toxikologie für Naturwissenschaftler und Mediziner; Stoffe, Mechanismen, Prüfverfahren, 2005, Wiley-VCH.
84. Eisenbrand, G., B. Pool-Zobel, V. Baker, M. Balls, B.J. Blaauboer, A. Boobis, A. Carere, S. Kevekordes, J.C. Lhuguenot, R. Pieters, and J. Kleiner, Methods of in vitro toxicology. Food and Chemical Toxicology, 2002. **40**(2-3): p. 193-236.
85. Verordnung über die Qualität von Wasser für den menschlichen Gebrauch. 2011.
86. Verordnung zum Schutz des Grundwassers: Grundwasserverordnung vom 9. November 2010 (BGBl. I S. 1513). 2010.
87. DIN EN ISO 10993-5:2009-10: Biologische Beurteilung von Medizinprodukten.
88. OSPAR, Survey on Genotoxicity Test Methods for the Evaluation of Waste Water within Whole Effluent Assessment. 2002.
89. Galloway, S.M., M.J. Aardema, M. Ishidate, J.L. Ivett, D.J. Kirkland, T. Morita, P. Mosesso, and T. Sofuni, Report from Working Group on In-vitro Tests for Chromosomal-Aberrations. Mutation Research, 1994. **312**(3): p. 241-261.
90. EPA, Health Effects Test Guidelines OPPTS 870.5375: In Vitro Mammalian Chromosome Abberation Test. 1998.
91. OECD, Test No. 473: Guideline for the Testing of Chemicals: In vitro Mammalian Chromosome Abberation Test. 1997.
92. OECD, Test No. 479: Genetic Toxicology: In Vitro Sister Chromatid Exchange Assay in Mammalian Cells. 1986.
93. Latt, S.A., J. Allen, S.E. Bloom, A. Carrano, E. Falke, D. Kram, E. Schneider, R. Schreck, R. Tice, B. Whitfield, and S. Wolff, Sister-Chromatid Echanges - A Report of the Gene-Tox Program. Mutation Research, 1981. **87**(1): p. 17-62.
94. OECD, Test No. 482: Genetic Toxicology: DNA Damage and Repair/Unscheduled DNA Synthesis in Mammalian Cells in vitro. 1986.
95. OECD, Test No. 473: OECD Guideline for the Testing of Chemicals - In Vitro Mammalian Cell Micronucleus Test. 2010.
96. Kirsch-Volders, M., A. Elhajouji, E. Cundari, and P. Van Hummelen, The in vitro micronucleus test: a multi-endpoint assay to detect simultaneously mitotic delay, apoptosis, chromosome breakage, chromosome loss and non-

References

disjunction. Mutation Research/Genetic Toxicology and Environmental Mutagenesis, 1997. **392**(1): p. 19-30.

97. Kohn, K.W. and Grimekew.Ra, *Alkaline Elution Analysis, a New Approach to the Study of DNA Single-Strand Interruptions in Cells.* Cancer Research, 1973. **33**(8): p. 1849-1853.

98. Storer, R.D., T.W. McKelvey, A.R. Kraynak, M.C. Elia, J.E. Barnum, L.S. Harmon, W.W. Nichols, and J.G. DeLuca, *Revalidation of the in vitro alkaline elution rat hepatocyte assay for DNA damage: Improved criteria for assessment of cytotoxicity and genotoxicity and results for 81 compounds.* Mutation Research-Genetic Toxicology, 1996. **368**(2): p. 59-101.

99. Swenberg, J.A., G.L. Petzold, and P.R. Harbach, *Invitro DNA Damage Alkaline Elution Assay for Predicting Carcinogenic Potential.* Biochemical and Biophysical Research Communications, 1976. **72**(2): p. 732-738.

100. BDS. *http://www.biodetectionsystems.com/1/news/76.php*. 2010.

101. Tice, R.R., E. Agurell, D. Anderson, B. Burlinson, A. Hartmann, H. Kobayashi, Y. Miyamae, E. Rojas, J.C. Ryu, and Y.F. Sasaki, *Single cell gel/comet assay: Guidelines for in vitro and in vivo genetic toxicology testing.* Environmental and Molecular Mutagenesis, 2000. **35**(3): p. 206-221.

102. Ostling, O. and K.J. Johanson, *Microelectrophoretic Study of Radiation-Induced DNA Damages in Individual Mammalian-Cells.* Biochemical and Biophysical Research Communications, 1984. **123**(1): p. 291-298.

103. Singh, N.P., M.T. McCoy, R.R. Tice, and E.L. Schneider, *A Simple Technique for Quantitation of Low-Levels of DNA Damage In Individual Cells.* Experimental Cell Research, 1988. **175**(1): p. 184-191.

104. Walmsley, R.M., *GADD45a-GFP GreenScreen HC genotoxicity screening assay.* Expert Opinion on Drug Metabolism & Toxicology, 2008. **4**(6): p. 827-835.

105. Aardema, M.J. and J.T. MacGregor, *Toxicology and genetic toxicology in the new era of "toxicogenomics": impact of "-omics" technologies.* Mutation Research-Fundamental and Molecular Mechanisms of Mutagenesis, 2002. **499**(1): p. 13-25.

106. Malik, A.I., A. Williams, C.L. Lemieux, P.A. White, and C.L. Yauk, *Hepatic mRNA, microRNA, and miR-34a-Target responses in mice after 28 days exposure to doses of benzo(a)pyrene that elicit DNA damage and mutation.* Environmental and Molecular Mutagenesis, 2012. **53**(1): p. 10-21.

107. Scott, D.J., A.S. Devonshire, Y.A. Adeleye, M.E. Schutte, M.R. Rodrigues, T.M. Wilkes, M.G. Sacco, L. Gribaldo, M. Fabbri, S. Coecke, M. Whelan, N. Skinner, A. Bennett, A. White, and C.A. Foy, *Inter- and intra-laboratory study to determine the reproducibility of toxicogenomics datasets.* Toxicology, 2011. **290**(1): p. 50-58.

108. OECD, *Test No: 476 Guideline for the Testing of Chemicals: In vitro Mammalian Cell Gene Mutation Test.* 1997.

109. EPA, *Health Effects Test Guidelines: OPPTS 870.5300: Detection of Gene Mutations in Somatic Cells in Culture.* 1996.

110. Clive, D. and J.F.S. Spector, *Laboratory Procedure for Assessing Specific Locus Mutations at TK Locus in Cultured L5178Y Mouse Lymphoma-Cells.* Mutation Research, 1975. **31**(1): p. 17-29.

111. OECD, *Test No: 480 Guideline for the Testing of Chemicals: genetic Toxicology: Saccharomyces cerevisiae, Gene Mutation Assay.* 1986.

References

112. OECD, *Test No: 481 Guideline for the Testing of Chemicals: Genetic Toxicology: Saccharomyces cerevisiae, Mitotic Recombination Assay.* 1986.
113. OECD, *Test No: 471 Guideline for the Testing Of Chemicals: Bacterial Reverse Mutation Test.* 1997.
114. FDA, http://www.fda.gov/Food/GuidanceComplianceRegulatoryInformation/GuidanceDocuments/FoodIngredientsandPackaging/Redbook/ucm078330.htm. 2000.
115. Ames, B.N., F.D. Lee, and W.E. Durston, *Improved Bacterial Test System for Detection and Classification of Mutagens and Carcinogens.* Proceedings of the National Academy of Sciences of the United States of America, 1973. **70**(3): p. 782-786.
116. Kleine, B., Rossmanith, W., *Hormone und Hormonsystem, Lehrbuch der Endokrinologie.* Springer, 2010. **2. Edition**.
117. Routledge, E.J., D. Sheahan, C. Desbrow, G.C. Brighty, M. Waldock, and J.P. Sumpter, *Identification of estrogenic chemicals in STW effluent. 2. In vivo responses in trout and roach.* Environmental Science & Technology, 1998. **32**(11): p. 1559-1565.
118. Kidd, K.A., P.J. Blanchfield, K.H. Mills, V.P. Palace, R.E. Evans, J.M. Lazorchak, and R.W. Flick, *Collapse of a fish population after exposure to a synthetic estrogen.* Proceedings of the National Academy of Sciences, 2007. **104**(21): p. 8897-8901.
119. Kramer, V.J., S. Miles-Richardson, S.L. Pierens, and J.P. Giesy, *Reproductive impairment and induction of alkaline-labile phosphate, a biomarker of estrogen exposure, in fathead minnows (Pimephales promelas) exposed to waterborne 17β-estradiol.* Aquatic Toxicology, 1998. **40**(4): p. 335-360.
120. Sheahan, S.A., D. Bucke, J.P. Matthiessen, J.P. Sumpter, M.F. Kirby, M. Neall, and M. Waldock, *The effects of low level 17-a ethynylestradiol upon plasma vitellogenin levels in male and female rainbow trout.* M. R. L. R. (Ed.). Sublethal and Chronic Effects of Pollutants on Freshwater Fish. FAD, Fishing News Books, Oxford, 1994.
121. Routledge, E.J. and J.P. Sumpter, *Estrogenic activity of surfactants and some of their degradation products assessed using a recombinant yeast screen.* Environmental Toxicology and Chemistry, 1996. **15**(3): p. 241-248.
122. Soto, A.M., C. Sonnenschein, K.L. Chung, M.F. Fernandez, N. Olea, and F.O. Serrano, *The E-Screen Assay as a Tool to Identify Estrogens - An Update on Estrogenic Environemntal-Pollutants.* Environmental Health Perspectives, 1995. **103**: p. 113-122.
123. Soto, A.M., K.L. Chung, and C. Sonnenschein, *The Pesticides Endosulfan, Toxaphene, and Dieldrin have Estrogenic Effects on Human Estrogen-Sensitive Cells.* Environmental Health Perspectives, 1994. **102**(4): p. 380-383.
124. Petit, F., Y. Valotaire, and F. Pakdel, *Differential Functional Activities of Rainbow-Trout and Human Estrogen-Receptors Expressed in the Yeast Saccharomyces-Cerevisiae.* European Journal of Biochemistry, 1995. **233**(2): p. 584-592.
125. Kunz, P.Y., H.F. Galicia, and K. Fent, *Comparison of in vitro and in vivo estrogenic activity of UV filters in fish.* Toxicological Sciences, 2006. **90**(2): p. 349-361.
126. Uhlig, S. and P. Gowik, *Factorial interlaboratory tests for microbiological measurement methods.* Journal für Verbraucherschutz und Lebensmittelsicherheit, 2010. **5**(1): p. 35-46.

References

127. Hahn, T., K. Tag, K. Riedel, S. Uhlig, K. Baronian, G. Gellissen, and G. Kunze, *A novel estrogen sensor based on recombinant Arxula adeninivorans cells.* Biosensors and Bioelectronics, 2006. **21**(11): p. 2078-2085.
128. Legler, J., van den Brink, C.E., Brouwer, A., Murk, A.J., van der Saag, P.T., Vethaak, A.D., van der Burg, B., *Development of a Stably Transfected Estrogen Receptor-Mediated Luciferase Reporter Gene Assay in the Human T47D Breast Cancer Cell Line.* Toxicological Science, 1999. **48**: p. 55-66.
129. OECD, *Test No. 455: Guideline for the testing of chemicals: Stably Transfected Human Estrogen-α Transcriptional Activation Assay for Detection of Estrogenic Agonist-Activity of Chemicals.* 2009.
130. Demirpence, E., M.-J. Duchesne, E. Badia, D. Gagne, and M. Pons, *MVLN Cells: A bioluminescent MCF-7-derived cell line to study the modulation of estrogenic activity.* The Journal of Steroid Biochemistry and Molecular Biology, 1993. **46**(3): p. 355-364.
131. Pons, M., D. Gagne, J.C. Nicolas, and M. Mehtali, *A new cellular model of response to estrogens: a bioluminescent test to characterize (anti) estrogen molecules.* BioTechniques, 1990. **9**(4): p. 450-9.
132. Ackermann, G.E., E. Brombacher, and K. Fent, *Development of a fish reporter gene system for the assessment of estrogenic compounds and sewage treatment plant effluents.* Environmental Toxicology and Chemistry, 2002. **21**(9): p. 1864-1875.
133. Gazdar, A.F., H.K. Oie, C.H. Shackleton, T.R. Chen, T.J. Triche, C.E. Myers, G.P. Chrousos, M.F. Brennan, C.A. Stein, and R.V. Larocca, *Establishment and Characterization of a Human Adrenocortical Carcinoma Cell-Line That Expresses Multiple Pathways of Steroid-Biosynthesis.* Cancer Research, 1990. **50**(17): p. 5488-5496.
134. OECD, *Test No. 456: Guideline for the testing of chemicals: H295R Steroidogenesis Assay.* 2011.
135. Seifert, M., S. Haindl, and B. Hock, *Development of an enzyme linked receptor assay (ELRA) for estrogens and xenoestrogens.* Analytica Chimica Acta, 1999. **386**(3): p. 191-199.
136. Huet, M.-C., *OECD Activity on Endocrine Disrupters Test Guidelines Development.* Ecotoxicology, 2000. **9**: p. 77-84.
137. OECD, *Conceptual Framework for the Testing and Assessment of Endocrine Disrupting Chemicals.* 2002.
138. Gelbke, H.P., M. Kayser, and A. Poole, *OECD test strategies and methods for endocrine disruptors.* Toxicology, 2004. **205**(1-2): p. 17-25.
139. Puck, T.T., Cieciura, S.J. and Robinson, A., *Genetics of somatic mammalian cells. III. Long-term cultivation of euploid cells from human and animal subjects.* J Exp Med, 1958. **108**(6): p. 945-956.
140. Schmitz, S., *Der Experimentator Zellkultur.* Spektrum Akademischer Verlag, 2009. **2. Auflage**: p. ISBN 978-3-8274-2108-1.
141. Repetto, G., A. del Peso, and J.L. Zurita, *Neutral red uptake assay for the estimation of cell viability/cytotoxicity.* Nature Protocols, 2008. **3**(7): p. 1125-1131.
142. Karp, G., *Molekulare Zellbiologie.* Springer, 2005. **4. Auflage**.
143. Skehan, P., R. Storeng, D. Scudiero, A. Monks, J. McMahon, D. Vistica, J.T. Warren, H. Bokesch, S. Kenney, and M.R. Boyd, *New Colorimetric Cytotoxicity Assay for Anticancer-Drug Screening.* Journal of the National Cancer Institute, 1990. **82**(13): p. 1107-1112.

References

144. Mosmann, T., *Rapid colorimetric assay for cellular growth and survival: Application to proliferation and cytotoxicity assays.* Journal of Immunological Methods, 1983. **65**(1-2): p. 55-63.
145. Gareis, M., *Diagnostischer Zellkulturtest (MTT-Test) für den Nachweis von zytotoxischen Kontaminanten und Rückständen.* Journal für Verbraucherschutz und Lebensmittelsicherheit, 2006. **1**(4): p. 354-363.
146. *DIN EN ISO 10993-5: Biologische Beurteilung von Medizinprodukten - Teil 5: Prüfungen auf In-vitro-Zytotoxizität* 2009.
147. Olive, P.L., J.P. Banáth, and R.E. Durand, *Heterogeneity in Radiation-Induced DNA Damage and Repair in Tumor and Normal Cells Measured Using the "Comet" Assay.* Radiation Research, 1990. **122**(1): p. 86-94.
148. Ames, B.N., J. McCann, and E. Yamasaki, *Methods for detecting carcinogens and mutagens with the salmonella/mammalian-microsome mutagenicity test.* Mutation Research/Environmental Mutagenesis and Related Subjects, 1975. **31**(6): p. 347-363.
149. Mortelmans, K. and E. Zeiger, *The Ames Salmonella/microsome mutagenicity assay.* Mutation Research/Fundamental and Molecular Mechanisms of Mutagenesis, 2000. **455**(1-2): p. 29-60.
150. BUND, *Hormonaktive Substanzen im Wasser - Gefahr für Gewässer und Mensch.* BUNDhintergrund, 2001.
151. *Abschlussbericht für das IGF-Forschungsvorhaben 15862 N: "Untersuchungen zur Bewertung und Vermeidung von toxischen Oxidationsnebenprodukten bei der oxidativen Abwasserbehandlung"* 2011.
152. Otto, M., *Analytische Chemie.* Wiley-VCH, 2000. **2. Edition**: p. ISBN 3-527-29840-1.
153. *Lexikon der Biochemie in zwei Teilen.* ELSEVIER Spektrum Akademischer Verlag, 2000. **ISBN: 3-8274-1580-2**.
154. Sanderson, J.E., S.K.W. Chan, G. Yip, L.Y.C. Yeung, K.W. Chan, K. Raymond, and K.S. Woo, *Beta-blockade in heart failure - A comparison of carvedilol with metoprolol.* Journal of the American College of Cardiology, 1999. **34**(5): p. 1522-1528.
155. Aktories, K., Förstermann, U., Hofmann, F.B., Starke, K., *Allgemeine und spezielle Pharmakologie und Toxikologie.* ELSEVIER Urban & Fischer, 2005. **9. Edition**: p. ISBN: 978-3-437-44490-6.
156. Swedenborg, E., J. Ruegg, S. Makela, and I. Pongratz, *Endocrine disruptive chemicals: mechanisms of action and involvement in metabolic disorders.* Journal of Molecular Endocrinology, 2009. **43**(1-2): p. 1-10.
157. Casals-Casas, C. and B. Desvergne, *Endocrine disruptors: from endocrine to metabolic disruption.* Annu Rev Physiol, 2011. **73**: p. 135-62.
158. Van der Linden, S.C., M.B. Heringa, H.Y. Man, E. Sonneveld, L.M. Puijker, A. Brouwer, and B. Van der Burg, *Detection of multiple hormonal activities in wastewater effluents and surface water, using a panel of steroid receptor CALUX bioassays.* Environmental Science & Technology, 2008. **42**(15): p. 5814-5820.
159. EHSC, *Note on: WHY DO WE WORRY ABOUT BISPHENOL-A.* http://www.rsc.org/images/bisphenol-a_tcm18-217036.pdf, 2012.
160. Burgis, E., *Intensivkurs Allgemeine und spezielle Pharmakologie* ELSEVIER Urban & Fischer, 2008. **4. Auflage**.
161. Kuse, M., Sandner F., *BASICS Allgemeine Pharmakologie.* ELSEVIER Urban & Fischer, 2009. **1. Auflage**.

References

162. Fleming, A., *On the specific antibacterial properties of penicillin and potassium tellurite - Incorporating a method of demonstrating some bacterial antiagonisms.* Journal of Pathology and Bacteriology, 1932. **35**(6): p. 831-842.
163. GERMAP 2010 - *Antibiotika-Resistenz und -Verbrauch.* Bundesamt für Verbaucherschutz und Lebensmittelsicherheit, 2011. **ISBN: 978-3-00-031622-7**(1. Auflage).
164. PubChem, *Sulfamethoxazole - Compound Summary* http://pubchem.ncbi.nlm.nih.gov/summary/summary.cgi?sid=602530&viewopt =PubChem&ncount=15#x94.
165. Power, E.G. and I. Phillips, *Induction of the SOS gene (umuC) by 4-quinolone antibacterial drugs.* J Med Microbiol, 1992. **36**(2): p. 78-82.
166. Mamber, S.W., B. Kolek, K.W. Brookshire, D.P. Bonner, and J. Fung-Tomc, *Activity of quinolones in the Ames Salmonella TA102 mutagenicity test and other bacterial genotoxicity assays.* Antimicrobial Agents and Chemotherapy, 1993. **37**(2): p. 213-217.
167. Hartmann, A., E.M. Golet, S. Gartiser, A.C. Alder, T. Koller, and R.M. Widmer, *Primary DNA damage but not mutagenicity correlates with ciprofloxacin concentrations in German hospital wastewaters.* Archives of Environmental Contamination and Toxicology, 1999. **36**(2): p. 115-119.
168. Prival, M.J. and E. Zeiger, *Chemicals mutagenic in Salmonella typhimurium strain TA1535 but not in TA100.* Mutation Research-Genetic Toxicology and Environmental Mutagenesis, 1998. **412**(3): p. 251-260.
169. Oliphant, C.M., Green, G.M., *Quinolones: a comprehensive review.* American Family Physician, 2002. **65**(3): p. 455-465.
170. Adolfsson-Erici, M., M. Pettersson, J. Parkkonen, and J. Sturve, *Triclosan, a commonly used bactericide found in human milk and in the aquatic environment in Sweden.* Chemosphere, 2002. **46**(9-10): p. 1485-1489.
171. Dann, A.B. and A. Hontela, *Triclosan: environmental exposure, toxicity and mechanisms of action.* Journal of Applied Toxicology, 2011. **31**(4): p. 285-311.
172. Bester, K., *Fate of triclosan and triclosan-methyl in sewage treatment plants and surface waters.* Archives of Environmental Contamination and Toxicology, 2005. **49**(1): p. 9-17.
173. Chen, X., J. Richard, Y. Liu, E. Dopp, J. Tuerk, and K. Bester, *Ozonation products of triclosan in advanced wastewater treatment.* Water Research, 2012. **46**(7): p. 2247-2256.
174. EU, *Richtlinie 98/8/EG des Europäischen Parlaments und des Rates vom 16. Februar 1998 über das Inverkehrbringen von Biozid-Produkten.* 1998.
175. Projekt SEA - Stoffdatenblatt Irgarol 1051. http://www.sea.eawag.ch/inhalt/sites/stoffe/pdf/Biozide_d.pdf.
176. Moretti, M., M. Marcarelli, M. Villarini, C. Fatigoni, G. Scassellati-Sforzolini, and R. Pasquini, *In vitro testing for genotoxicity of the herbicide terbutryn: cytogenetic and primary DNA damage.* Toxicology in Vitro, 2002. **16**(1): p. 81-88.
177. Broser, M., C. Gloeckner, A. Gabdulkhakov, A. Guskov, J. Buchta, J. Kern, F. Mueh, H. Dau, W. Saenger, and A. Zouni, *Structural basis of cyanobacterial photosystem II inhibition by the herbicide terbutryn.* Journal of Biological Chemistry, 2011.
178. Nilson, E.L. and R.F. Unz, *Antialgal substances for iodine-disinfected swimming pools.* Applied and Environmental Microbiology, 1977. **34**(6): p. 815-822.

References

179. HERA, *Polycyclic musks AHTN (CAS 1506-02-1) and HHCB (CAS 1222-05-05)*. 2004.
180. ECHA, *Candidate List of Substances of Very High Concern for Authorisation.* http://echa.europa.eu/web/guest/candidate-list-table, 2011.
181. Rimkus, G.G., *Synthetic musk fragrances as a new class of environmental pollutants.* Symposia Papers presented before the Division of Environmental Chemistry, Americal Chemical Society, 1999.
182. Uhl, M., Hutter, H.P. Lorbeer, G., *Polymuschusverbindungen in Humanblut II: Humanbiomonitoring von Moschusduftstoffen.* Final Report for the Austrian Ministry of Health 2005.
183. Kupper, T., C. Plagellat, R.C. Braendli, L.F. de Alencastro, D. Grandjean, and J. Tarradellas, *Fate and removal of polycyclic musks, UV filters and biocides during wastewater treatment.* Water Research, 2006. **40**(14): p. 2603-2612.
184. Simonich, S.L., T.W. Federle, W.S. Eckhoff, A. Rottiers, S. Webb, D. Sabaliunas, and W. De Wolf, *Removal of fragrance materials during US and European wastewater treatment.* Environmental Science & Technology, 2002. **36**(13): p. 2839-2847.
185. EC, *SCHER, scientific opinion on the risk assessment report on 1-(5,6,7,8-tetrahydro-3.5.56.8.8-hexamethyl-2-naphtyl)ethan-1-one (AHTN), human health part, 20 September 2007.* 2007.
186. Janzen, N., E. Dopp, J. Hesse, J. Richards, J. Türk, and K. Bester, *Transformation products and reaction kinetics of fragrances in advanced wastewater treatment with ozone.* Chemosphere, 2011. **85**(9): p. 1481-1486.
187. Sagunski, H. and E. Roßkamp, *Richtwerte üfr die Innenraumluft: Tris(2-chloroethyl)phosphat.* Budensgesundheitsbl - Gesundheitsforsch - Gesundheitsschutz, 2002. **45**: p. 300-306.
188. Raut, S.A. and R.A. Angus, *Triclosan has Endocrine-Disrupting Effects in Male Western Mosquitofish, Gambusia affinis.* Environmental Toxicology and Chemistry, 2010. **29**(6): p. 1287-1291.
189. Sumpter, J.P., *Endocrine disrupters in the aquatic environment: An overview.* Acta Hydrochimica Et Hydrobiologica, 2005. **33**(1): p. 9-16.
190. Johnson, A. and M. Jurgens, *Endocrine active industrial chemicals: Release and occurrence in the environment.* Pure and Applied Chemistry, 2003. **75**(11-12): p. 1895-1904.
191. Crisp, T.M., Clegg, E.D., Coper, R.L., Anderson, D.G., Baetcke, K.P., Hoffmann, J.L., Morrow, M.S., Rodier, D.J., Schaeffer, J.E., Touart, L.W., Zeeman, M.G., Patel, Y.M., Wood, W.P., *Special Report on Environmental Endocrine Disruption: An Effects Assessment and Analysis.* EPA/630/R-96/012, 1997.
192. Lintelmann, J., A. Katayama, N. Kurihara, L. Shore, and A. Wenzel, *Endocrine disruptors in the environment - (IUPAC Technical Report).* Pure and Applied Chemistry, 2003. **75**(5): p. 631-681.
193. BKH, *European Commission DG ENV. Towards the establishment of a priority list of substances for further evaluation of their role in endocrine disruption - preparation of a candidate list of substances as a basis for priority setting. Final Report. BKH Consulting Engineers , Delft, The Netherlands in association woth TNO nutrition and Food Research, Zeist, The Netherlands.* 2000.
194. Jobling, S., R. Williams, A. Johnson, A. Taylor, M. Gross-Sorokin, M. Nolan, C.R. Tyler, R. van Aerle, E. Santos, and G. Brighty, *Predicted exposures to*

References

steroid estrogens in UK rivers correlate with widespread sexual disruption in wild fish populations. Environmental Health Perspectives, 2006. **114**: p. 32-39.

195. Jobling, S., M. Nolan, C.R. Tyler, G. Brighty, and J.P. Sumpter, *Widespread sexual disruption in wild fish.* Environmental Science & Technology, 1998. **32**(17): p. 2498-2506.

196. Guillette, L.J., D.A. Crain, M.P. Gunderson, S.A.E. Kools, M.R. Milnes, E.F. Orlando, A.A. Rooney, and A.R. Woodward, *Alligators and endocrine disrupting contaminants: A current perspective.* American Zoologist, 2000. **40**(3): p. 438-452.

197. Liney, K.E., J.A. Hagger, C.R. Tyler, M.H. Depledge, T.S. Galloway, and S. Jobling, *Health effects in fish of long-term exposure to effluents from wastewater treatment works.* Environmental Health Perspectives, 2006. **114**: p. 81-89.

198. Giesy, J.P., L.A. Feyk, P.D. Jones, K. Kannan, and T. Sanderson, *Review of the effects of endocrine-disrupting chemicals in birds.* Pure and Applied Chemistry, 2003. **75**(11-12): p. 2287-2303.

199. Newbold, R.R., E. Padilla-Banks, W.N. Jefferson, and J.J. Heindel, *Effects of endocrine disruptors on obesity.* International Journal of Andrology, 2008. **31**(2): p. 201-207.

200. Newbold, R.R., E. Padilla-Banks, R.J. Snyder, and W.N. Jefferson, *Perinatal exposure to environmental estrogens and the development of obesity.* Molecular Nutrition & Food Research, 2007. **51**(7): p. 912-917.

201. Davis, D.L., H.L. Bradlow, M. Wolff, T. Woodruff, D.G. Hoel, and H. Antonculver, *Medical Hypothesis - Xenoestrogens as Preventable Causes of Breast-Cancer.* Environmental Health Perspectives, 1993. **101**(5): p. 372-377.

202. Donna, A., P.G. Betta, F. Robutti, P. Crosignani, F. Berrino, and D. Bellingeri, *Ovarian Mesothelial Tumors and Herbicieds - A Case-Control Study.* Carcinogenesis, 1984. **5**(7): p. 941-942.

203. Donna, A., *CORRECTION.* Carcinogenesis, 1984. **5**(10): p. 1380-1380.

204. Carpenter, D.O., *Human health effects of environmental pollutants: New insights.* Environmental Monitoring and Assessment, 1998. **53**(1): p. 245-258.

205. Melnick, R., G. Lucier, M. Wolfe, R. Hall, G. Stancel, G. Prins, M. Gallo, K. Reuhl, S.M. Ho, T. Brown, J. Moore, J. Leakey, J. Haseman, and M. Kohn, *Summary of the National Toxicology Program's report of the endocrine disruptors low-dose peer review.* Environ Health Perspect, 2002. **110**(4): p. 427-31.

206. Vandenberg, L.N., T. Colborn, T.B. Hayes, J.J. Heindel, D.R. Jacobs, D.H. Lee, T. Shioda, A.M. Soto, F.S. vom Saal, W.V. Welshons, R.T. Zoeller, and J.P. Myers, *Hormones and Endocrine-Disrupting Chemicals: Low-Dose Effects and Nonmonotonic Dose Responses.* Endocrine Reviews, 2012. **33**(3): p. 378-455.

207. Andersen, H.R., A.M. Vinggaard, T.H. Rasmussen, I.M. Gjermandsen, and E.C. Bonefeld-Jorgensen, *Effects of currently used pesticides in assays for estrogenicity, androgenicity, and aromatase activity in vitro.* Toxicology and Applied Pharmacology, 2002. **179**(1): p. 1-12.

208. Kinnberg, K., *Evaluation of in vitro assays for determination of estrogenic activity in the environment.* Working Report No. 43, Danish Environmental Protection Agency, 2003.

209. Shelby, M.D., R.R. Newbold, D.B. Tully, K. Chae, and V.L. Davis, *Assessing environmental chemicals for estrogenicity using a combination of in vitro and*

in vivo assays. Environmental Health Perspectives, 1996. **104**(12): p. 1296-1300.
210. Shanle, E.K. and W. Xu, *Endocrine Disrupting Chemicals Targeting Estrogen Receptor Signaling: Identification and Mechanisms of Action.* Chemical Research in Toxicology, 2011. **24**(1): p. 6-19.
211. Giger, W., Alder, A.C., Golet, E. M., Kohler, H.-P.E., McArdell, C. S., Molnar, E., Siegrist, H., Suter, M.J.F., *Occurrence and Fate of Antibiotics as Trace Contaminants in Wastewaters, Sewage Sludges, and Surface Waters* CHIMIA International Journal for Chemistry, 2003. **9**(9): p. 485-491.
212. BMG, *Budesministerium für Gesundheit: DART German Antimicrobial Resistance Strategy.* http://www.bmelv.de/SharedDocs/Downloads/EN/Agriculture/GermanAntimicrobialResistanceStrategy.pdf?__blob=publicationFile, 2008.
213. WHO, *World Health Organization, Department of essential drugs and medicines policy: WHO Workshop on containment of antimicrobial resistance in Europe, 26–27 February 2004 in Wernigerode, Germany.* Bundesgesundheitsbl - Gesundheitsforsch - Gesundheitsschutz, 2005. **48**: p. 221-131.
214. Jevons, M.P., G.N. Rolinson, and R. Knox, *CELBENIN-RESISTANT STAPHYLOCOCCI.* British Medical Journal, 1961. **1**(521): p. 124-&.
215. Livermore, D.M., *Antibiotic resistance in staphylococci.* International Journal of Antimicrobial Agents, 2000. **16**: p. S3-S10.
216. Madigan, M.T., J.M. Martinko, and J. Parker, *Brock Mikrobiologie*, ed. G. W.2002, Berlin: Spektrum Akademischer Verlag Heidelberg Berlin.
217. Dodd, M.C., H.-P.E. Kohler, and U. Von Gunten, *Oxidation of Antibacterial Compounds by Ozone and Hydroxyl Radical: Elimination of Biological Activity during Aqueous Ozonation Processes.* Environmental Science & Technology, 2009. **43**(7): p. 2498-2504.
218. vom Eyser, C., *Degradation of Fluoroquinolones Ofloxacin and Ciprofloxacin via Ozonation and UV-Oxidation.* Master Thesis, 2011.
219. Altenburger, R., T. Backhaus, W. Boedeker, M. Faust, M. Scholze, and L.H. Grimme, *Predictability of the toxicity of multiple chemical mixtures to Vibrio fischeri: Mixtures composed of similarly acting chemicals.* Environmental Toxicology and Chemistry, 2000. **19**(9): p. 2341-2347.
220. Porsbring, T., T. Backhaus, P. Johansson, M. Kuylenstierna, and H. Blanck, *Mixture Toxicity from Photosystem II Inhibitors on Microalgal Community Succession is Predictable by Concentration Addition.* Environmental Toxicology and Chemistry, 2010. **29**(12): p. 2806-2813.
221. Lin, K., *Joint acute toxicity of tributyl phosphate and triphenyl phosphate to Daphnia magna.* Environmental Chemistry Letters, 2009. **7**(4): p. 309-312.
222. Chevre, N., C. Loeppe, H. Singer, C. Stamm, K. Fenner, and B.I. Escher, *Including mixtures in the determination of water quality criteria for herbicides in surface water.* Environmental Science & Technology, 2006. **40**(2): p. 426-435.
223. Brian, J.V., C.A. Harris, M. Scholze, T. Backhaus, P. Booy, M. Lamoree, G. Pojana, N. Jonkers, T. Runnalls, A. Bonfa, A. Marcomini, and J.P. Sumpter, *Accurate prediction of the response of freshwater fish to a mixture of estrogenic chemicals.* Environmental Health Perspectives, 2005. **113**(6): p. 721-728.

References

224. Altenburger, R., H. Walter, and M. Grote, *What contributes to the combined effect of a complex mixture?* Environmental Science & Technology, 2004. **38**(23): p. 6353-6362.
225. Escher, B.I. and J.L.M. Hermens, *Modes of action in ecotoxicology: Their role in body burdens, species sensitivity, QSARs, and mixture effects.* Environmental Science & Technology, 2002. **36**(20): p. 4201-4217.
226. Kortenkamp, A. and R. Altenburger, *Approaches to assessing combination effects of oestrogenic environmental pollutants.* Science of the Total Environment, 1999. **233**(1-3): p. 131-140.
227. Hernando, M.D., M. Ejerhoon, A.R. Fernandez-Alba, and Y. Chisti, *Combined toxicity effects of MTBE and pesticides measured with Vibrio fischeri and Daphnia magna bioassays.* Water Research, 2003. **37**(17): p. 4091-4098.
228. Fernandez-Alba, A.R., M.D. Hernando, L. Piedra, and Y. Chisti, *Toxicity evaluation of single and mixed antifouling biocides measured with acute toxicity bioassays.* Analytica Chimica Acta, 2002. **456**(2): p. 303-312.
229. Dourson, M.L. and J.F. Stara, *Regulatory history and experimental support of uncertainty (safety) factors.* Regulatory Toxicology and Pharmacology, 1983. **3**(3): p. 224-238.
230. Fairhurst, S., *The Uncertainty Factor in the Setting of Occupational Exposure Standards.* Annals of Occupational Hygiene, 1995. **39**(3): p. 375-385.
231. Xiong, J., *Essential Bioinformatics* 2007, New York: Cambridge University Press.
232. Escher, B.I., N. Bramaz, R.I.L. Eggen, and M. Richter, *In vitro assessment of modes of toxic action of pharmaceuticals in aquatic life.* Environmental Science & Technology, 2005. **39**(9): p. 3090-3100.
233. Escher, B.I., W. Pronk, M.J.F. Suter, and M. Maurer, *Monitoring the removal efficiency of pharmaceuticals and hormones in different treatment processes of source-separated urine with bioassays.* Environmental Science & Technology, 2006. **40**(16): p. 5095-5101.
234. Brack, W., *Effect-directed analysis: a promising tool for the identification of organic toxicants in complex mixtures?* Analytical and Bioanalytical Chemistry, 2003. **377**(3): p. 397-407.
235. Hecker, M. and H. Hollert, *Effect-directed analysis (EDA) in aquatic ecotoxicology: state of the art and future challenges.* Environmental Science and Pollution Research, 2009. **16**(6): p. 607-613.
236. Ankley, G.T., R.S. Bennett, R.J. Erickson, D.J. Hoff, M.W. Hornung, R.D. Johnson, D.R. Mount, J.W. Nichols, C.L. Russom, P.K. Schmieder, J.A. Serrrano, J.E. Tietge, and D.L. Villeneuve, *Adverse Outcome Pathways: A Conceptual Framework to Support Ecotoxicological Research and Risk Assessment.* Environmental Toxicology and Chemistry, 2010. **29**(3): p. 730-741.
237. Escher, B., Jose Farre, M., Neale, P., *Risk-oriented screening of disinfection by-products.* Water21, 2012(August 2012).
238. Stasinakis, A.S., G. Gatidou, D. Mamais, N.S. Thomaidis, and T.D. Lekkas, *Occurrence and fate of endocrine disrupters in Greek sewage treatment plants.* Water Research, 2008. **42**(6-7): p. 1796-1804.
239. Joss, A., H. Siegrist, and T.A. Ternes, *Are we about to upgrade wastewater treatment for removing organic micropollutants?* Water Science and Technology, 2008. **57**(2): p. 251-255.

References

240. Stalter, D., A. Magdeburg, M. Wagner, and J. Oehlmann, *Ozonation and activated carbon treatment of sewage effluents: Removal of endocrine activity and cytotoxicity.* Water Research, 2011. **45**(3): p. 1015-1024.
241. Guzzella, L., D. Feretti, and S. Monarca, *Advanced oxidation and adsorption technologies for organic micropollutant removal from lake water used as drinking-water supply.* Water Research, 2002. **36**(17): p. 4307-4318.
242. Hernando, M.D., S. Vettori, M.J.M. Bueno, and A.R. Fernandez-Alba, *Toxicity evaluation with Vibrio fischeri test of organic chemicals used in aquaculture.* Chemosphere, 2007. **68**(4): p. 724-730.
243. Farre, M. and D. Barcelo, *Toxicity testing of wastewater and sewage sludge by biosensors, bioassays and chemical analysis.* Trac-Trends in Analytical Chemistry, 2003. **22**(5): p. 299-310.
244. Schrank, S.G., U. Bieling, H.J. Jose, R. Moreira, and H.F. Schroder, *Generation of endocrine disruptor compounds during ozone treatment of tannery wastewater confirmed by biological effect analysis and substance specific analysis.* Water Science and Technology, 2009. **59**(1): p. 31-38.
245. Dirany, A., S.E. Aaron, N. Oturan, I. Sires, M.A. Oturan, and J.J. Aaron, *Study of the toxicity of sulfamethoxazole and its degradation products in water by a bioluminescence method during application of the electro-Fenton treatment.* Analytical and Bioanalytical Chemistry, 2011. **400**(2): p. 353-360.
246. Trovo, A.G., R.F.P. Nogueira, A. Aguera, A.R. Fernandez-Alba, and S. Malato, *Degradation of the antibiotic amoxicillin by photo-Fenton process - Chemical and toxicological assessment.* Water Research, 2011. **45**(3): p. 1394-1402.
247. Desbrow, C., E.J. Routledge, G.C. Brighty, J.P. Sumpter, and M. Waldock, *Identification of estrogenic chemicals in STW effluent. 1. Chemical fractionation and in vitro biological screening.* Environmental Science & Technology, 1998. **32**(11): p. 1549-1558.
248. Erger, C., P. Balsaa, F. Werres, and T.C. Schmidt, *Occurrence of residual water within disk-based solid-phase extraction and its effect on GC-MS measurement of organic extracts of environmental samples.* Analytical and Bioanalytical Chemistry, 2012. **403**(9): p. 2541-2552.
249. Zabiegala, B., A. Kot-Wasik, M. Urbanowicz, and J. Namiesnik, *Passive sampling as a tool for obtaining reliable analytical information in environmental quality monitoring.* Analytical and Bioanalytical Chemistry, 2010. **396**(1): p. 273-296.
250. Gorecki, T. and J. Namiesnik, *Passive sampling.* Trac-Trends in Analytical Chemistry, 2002. **21**(4): p. 276-291.
251. Seethapathy, S., T. Gorecki, and X.J. Li, *Passive sampling in environmental analysis.* Journal of Chromatography A, 2008. **1184**(1-2): p. 234-253.
252. Chen, C.E., H. Zhang, and K.C. Jones, *A novel passive water sampler for in situ sampling of antibiotics.* Journal of Environmental Monitoring, 2012. **14**(6): p. 1523-1530.
253. Vystavna, Y., F. Huneau, V. Grynenko, Y. Vergeles, H. Celle-Jeanton, N. Tapie, H. Budzinski, and P. Le Coustumer, *Pharmaceuticals in Rivers of Two Regions with Contrasted Socio-Economic Conditions: Occurrence, Accumulation, and Comparison for Ukraine and France.* Water Air and Soil Pollution, 2012. **223**(5): p. 2111-2124.
254. Cernoch, I., M. Franek, I. Diblikova, K. Hilscherova, T. Randak, T. Ocelka, and L. Blaha, *POCIS sampling in combination with ELISA: Screening of*

References

sulfonamide residues in surface and waste waters. Journal of Environmental Monitoring, 2012. **14**(1): p. 250-257.

255. Vermeirssen, E.L.M., O. Korner, R. Schonenberger, M.J.F. Suter, and P. Burkhardt-Holm, *Characterization of environmental estrogens in river water using a three pronged approach: Active and passive water sampling and the analysis of accumulated estrogens in the bile of caged fish.* Environmental Science & Technology, 2005. **39**(21): p. 8191-8198.

256. Lacey, C., S. Basha, A. Morrissey, and J.M. Tobin, *Occurrence of pharmaceutical compounds in wastewater process streams in Dublin, Ireland.* Environmental Monitoring and Assessment, 2012. **184**(2): p. 1049-1062.

257. Fraysse, B. and J. Garric, *Prediction and experimental validation of acute toxicity of beta-blockers in Ceriodaphnia dubia.* Environmental Toxicology and Chemistry, 2005. **24**(10): p. 2470-2476.

258. Cheong, H.I., J. Johnson, M. Cormier, and K. Hosseini, *In vitro cytotoxicity of eight beta-blockers in human corneal epithelial and retinal pigment epithelial cell lines: Comparison with epidermal keratinocytes and dermal fibroblasts.* Toxicology in Vitro, 2008. **22**(4): p. 1070-1076.

259. Telez, M., B. Martinez, B. Criado, C.M. Lostao, O. Penagarikano, B. Ortega, P. Flores, E. Ortiz-Lastra, R.M. Alonso, R.M. Jimenez, and I. Arrieta, *In vitro and in vivo evaluation of the antihypertensive drug atenolol in cultured human lymphocytes: effects of long-term therapy.* Mutagenesis, 2000. **15**(3): p. 195-202.

260. Parolini, M., B. Quinn, A. Binelli, and A. Provini, *Cytotoxicity assessment of four pharmaceutical compounds on the zebra mussel (Dreissena polymorpha) haemocytes, gill and digestive gland primary cell cultures.* Chemosphere, 2011. **84**(1): p. 91-100.

261. Ioannou, L.A., E. Hapeshi, M.I. Vasquez, D. Mantzavinos, and D. Fatta-Kassinos, *Solar/TiO(2) photocatalytic decomposition of beta-blockers atenolol and propranolol in water and wastewater.* Solar Energy, 2011. **85**(9): p. 1915-1926.

262. Huggett, D.B., B.W. Brooks, B. Peterson, C.M. Foran, and D. Schlenk, *Toxicity of select beta adrenergic receptor-blocking pharmaceuticals (B-blockers) on aquatic organisms.* Archives of Environmental Contamination and Toxicology, 2002. **43**(2): p. 229-235.

263. van den Brandhof, E.J. and M. Montforts, *Fish embryo toxicity of carbamazepine, diclofenac and metoprolol.* Ecotoxicology and Environmental Safety, 2010. **73**(8): p. 1862-1866.

264. Hernando, M.D., M. Petrovic, A.R. Fernandez-Alba, and D. Barcelo, *Analysis by liquid chromatography-electro spray ionization tandem mass spectrometry and acute toxicity evaluation for beta-blockers and lipid-regulating agents in wastewater samples.* Journal of Chromatography A, 2004. **1046**(1-2): p. 133-140.

265. Dzialowski, E.M., P.K. Turner, and B.W. Brooks, *Physiological and reproductive effects of beta adrenergic receptor antagonists in Daphnia magna.* Archives of Environmental Contamination and Toxicology, 2006. **50**(4): p. 503-510.

266. Maurer, M., B.I. Escher, P. Richle, C. Schaffner, and A.C. Alder, *Elimination of beta-blockers in sewage treatment plants.* Water Research, 2007. **41**(7): p. 1614-1622.

References

267. Benner, J. and T.A. Ternes, *Ozonation of Metoprolol: Elucidation of Oxidation Pathways and Major Oxidation Products.* Environmental Science & Technology, 2009. **43**(14): p. 5472-5480.
268. Vethaak, A.D., J. Lahr, S.M. Schrap, A.C. Belfroid, G.B.J. Rijs, A. Gerritsen, J. de Boer, A.S. Bulder, G.C.M. Grinwis, R.V. Kuiper, J. Legler, T.A.J. Murk, W. Peijnenburg, H.J.M. Verhaar, and P. de Voogt, *An integrated assessment of estrogenic contamination and biological effects in the aquatic environment of The Netherlands.* Chemosphere, 2005. **59**(4): p. 511-524.
269. Dussault, E.B., V.K. Balakrishnan, K.R. Solomon, and P.K. Sibley, *Chronic Toxicity of the Synthetic Hormone 17 alpha-Ethinylestradiol to Chironomus tentans and Hyalella azteca.* Environmental Toxicology and Chemistry, 2008. **27**(12): p. 2521-2529.
270. Murk, A.J., J. Legler, M.M.H. van Lipzig, J.H.N. Meerman, A.C. Belfroid, A. Spenkelink, B. van der Burg, G.B.J. Rijs, and D. Vethaak, *Detection of estrogenic potency in wastewater and surface water with three in vitro bioassays.* Environmental Toxicology and Chemistry, 2002. **21**(1): p. 16-23.
271. Lang, R. and R. Reimann, *Studies for a Genotoxicity Potential of Some Endogenous and Exogebous Sex Steroids. 1. Communication - Examination for the Induction of Gene-Mutations Using the Ames Salmonella Microsome Test and the HGPRT Test in V79 Cells.* Environmental and Molecular Mutagenesis, 1993. **21**(3): p. 272-304.
272. Reimann, R., S. Kalweit, and R. Lang, *Studies for a genotoxic potential of some endogenous and exogenous sex steroids .2. Communication: Examination for the induction of cytogenetic damage using the chromosomal aberration assay on human lymphocytes in vitro and the mouse bone marrow micronucleus test in vivo.* Environmental and Molecular Mutagenesis, 1996. **28**(2): p. 133-144.
273. Siddique, Y.H., T. Beg, and M. Afzal, *Genotoxic potential of ethinylestradiol in cultured mammalian cells.* Chemico-Biological Interactions, 2005. **151**(2): p. 133-141.
274. Hashimoto, S., E. Watanabe, M. Ikeda, Y. Terao, C.A. Struessmann, M. Inoue, and A. Hara, *Effects of Ethinylestradiol on Medaka (Oryzias latipes) as Measured by Sperm Motility and Fertilization Success.* Archives of Environmental Contamination and Toxicology, 2009. **56**(2): p. 253-259.
275. Salierno, J.D. and A.S. Kane, *17 alpha-Ethinylestradiol Alters Reproductive Behaviors, Circulating Hormones, and Sexual Morphology in Male Fathead Minnows (Pimephales promelas).* Environmental Toxicology and Chemistry, 2009. **28**(5): p. 953-961.
276. Caldwell, D.J., F. Mastrocco, T.H. Hutchinson, R. Lange, D. Heijerick, C. Janssen, P.D. Anderson, and J.P. Sumpter, *Derivation of an aquatic predicted no-effect concentration for the synthetic hormone, 17 alpha-ethinyl estradiol.* Environmental Science & Technology, 2008. **42**(19): p. 7046-7054.
277. Commission, E., *Commission Directive 2011/8/EU of 28 January 2011 amending Directive 2002/72/EC as regards the restriction of use of Bisphenol A in plastic infant feeding bottles.* 2011.
278. Metcalfe, C.D., T.L. Metcalfe, Y. Kiparissis, B.G. Koenig, C. Khan, R.J. Hughes, T.R. Croley, R.E. March, and T. Potter, *Estrogenic potency of chemicals detected in sewage treatment plant effluents as determined by in vivo assays with Japanese medaka (Oryzias latipes).* Environmental Toxicology and Chemistry, 2001. **20**(2): p. 297-308.

References

279. Wu, M.H., H. Xu, M. Yang, and G. Xu, *Effects of chronic bisphenol A exposure on hepatic antioxidant parameters in medaka (Oryzias latipes).* Toxicological and Environmental Chemistry, 2011. **93**(2): p. 270-278.
280. Zhang, H., F.X. Kong, Y. Yu, X.L. Shi, M. Zhang, and H.E. Tian, *Assessing the combination effects of environmental estrogens in fish.* Ecotoxicology, 2010. **19**(8): p. 1476-1486.
281. Golub, M.S., K.L. Wu, F.L. Kaufman, L.H. Li, F. Moran-Messen, L. Zeise, G.V. Alexeeff, and J.M. Donald, *Bisphenol A: Developmental Toxicity from Early Prenatal Exposure.* Birth Defects Research Part B-Developmental and Reproductive Toxicology, 2010. **89**(6): p. 441-466.
282. ECHA, *Community Rolling Action Plan (CoRAP).* 2012.
283. ECHA, http://apps.echa.europa.eu/registered/data/dossiers/DISS-9dbe071c-c12d-0fe1-e044-00144f67d249/DISS-9dbe071c-c12d-0fe1-e044-00144f67d249_DISS-9dbe071c-c12d-0fe1-e044-00144f67d249.html. Bisphenol A.
284. Sotelo, J.L., F.J. Beltran, F.J. Benitez, and J. Beltranheredia, *HENRYS LAW CONSTANT FOR THE OZONE WATER-SYSTEM.* Water Research, 1989. **23**(10): p. 1239-1246.
285. Ternes, T., *Assessment of Technologies for the Removal of Pharmaceuticals and Personal Care Products in Sewage and Drinking Water Facilities to Improve the Indirect Potable Water Reuse.* Final Report POSEIDON Project, 2006.
286. Abellan, M.N., W. Gebhardt, and H.F. Schroder, *Detection and identification of degradation products of sulfamethoxazole by means of LC/MS and -MSn after ozone treatment.* Water Science and Technology, 2008. **58**(9): p. 1803-1812.
287. Lam, M.W. and S.A. Mabury, *Photodegradation of the pharmaceuticals atorvastatin, carbamazepine, levofloxacin, and sulfamethoxazole in natural waters.* Aquatic Sciences, 2005. **67**(2): p. 177-188.
288. Yargeau, V., J.C. Huot, A. Rodayan, L. Rouleau, R. Roy, and R.L. Leask, *Impact of degradation products of sulfamethoxazole on mammalian cultured cells.* Environmental Toxicology, 2008. **23**(4): p. 492-498.
289. Baran, W., J. Sochacka, and W. Wardas, *Toxicity and biodegradability of sulfonamides and products of their photocatalytic degradation in aqueous solutions.* Chemosphere, 2006. **65**(8): p. 1295-1299.
290. del Mar Gomez-Ramos, M., M. Mezcua, A. Agueera, A.R. Fernandez-Alba, S. Gonzalo, A. Rodriguez, and R. Rosal, *Chemical and toxicological evolution of the antibiotic sulfamethoxazole under ozone treatment in water solution.* Journal of Hazardous Materials, 2011. **192**(1): p. 18-25.
291. Dantas, R.F., S. Contreras, C. Sans, and S. Esplugas, *Sulfamethoxazole abatement by means of ozonation.* Journal of Hazardous Materials, 2008. **150**(3): p. 790-794.
292. Escher, B.I., N. Bramaz, M. Maurer, M. Richter, D. Sutter, C. von Kanel, and M. Zschokke, *Screening test battery for pharmaceuticals in urine and wastewater.* Environmental Toxicology and Chemistry, 2005. **24**(3): p. 750-758.
293. Hartmann, A., A.C. Alder, T. Koller, and R.M. Widmer, *Identification of fluoroquinolone antibiotics as the main source of umuC genotoxicity in native hospital wastewater.* Environmental Toxicology and Chemistry, 1998. **17**(3): p. 377-382.

294. Gurbay, A., M. Osman, A. Favier, and F. Hincal, *Ciprofloxacin-induced cytotoxicity and apoptosis in HeLa cells.* Toxicology Mechanisms and Methods, 2005. **15**(5): p. 339-342.
295. Sousa, M.C. and J. Poiares-da-Silva, *The cytotoxic effects of ciprofloxacin in Giardia lamblia trophozoites.* Toxicology in Vitro, 2001. **15**(4-5): p. 297-301.
296. Clerch, B., J.M. Bravo, and M. Llagostera, *Analysis of the ciprofloxacin-induced mutations in Salmonella typhimurium.* Environmental and Molecular Mutagenesis, 1996. **27**(2): p. 110-115.
297. Ferrari, B., R. Mons, B. Vollat, B. Fraysse, N. Paxeus, R. Lo Giudice, A. Pollio, and J. Garric, *Environmental risk assessment of six human pharmaceuticals: Are the current environmental risk assessment procedures sufficient for the protection of the aquatic environment?* Environmental Toxicology and Chemistry, 2004. **23**(5): p. 1344-1354.
298. Bezwada, P., L.A. Clark, and S. Schneider, *Intrinsic cytotoxic effects of fluoroquinolones on human corneal keratocytes and endothelial cells.* Current Medical Research and Opinion, 2008. **24**(2): p. 419-424.
299. Tsai, T.H., W.L. Chen, and F.R. Hu, *Comparison of fluoroquinolones: cytotoxicity on human corneal epithelial cells.* Eye, 2010. **24**(5): p. 909-917.
300. Scuderi, A.C., G.M. Paladino, C. Marino, and F. Trombetta, *In vitro toxicity of netilmicin and ofloxacin on corneal epithelial cells.* Cornea, 2003. **22**(5): p. 468-472.
301. Trisciuoglio, D., E. Krasnowska, A. Maggi, R. Pozzi, T. Parasassi, and O. Sapora, *Phototoxic effect of fluoroquinolones on two human cell lines.* Toxicology in Vitro, 2002. **16**(4): p. 449-456.
302. Itoh, T., K. Mitsumori, S. Kawaguchi, and Y.F. Sasaki, *Genotoxic potential of quinolone antimicrobials in the in vitro comet assay and micronucleus test.* Mutation Research-Genetic Toxicology and Environmental Mutagenesis, 2006. **603**(2): p. 135-144.
303. McQueen, C.A., B.M. Way, S.M. Queener, G. Schluter, and G.M. Williams, *Study of Potential Invitro and Invivo Genotoxicity in Hepatocytes of Quinolone Antibiotics.* Toxicology and Applied Pharmacology, 1991. **111**(2): p. 255-262.
304. ECHA, http://apps.echa.europa.eu/registered/data/dossiers/DISS-9ea3b5cc-80fb-15ea-e044-00144f67d031/DISS-9ea3b5cc-80fb-15ea-e044-00144f67d031_DISS-9ea3b5cc-80fb-15ea-e044-00144f67d031.html. Triclosan.
305. Bedoux, G., B. Roig, O. Thomas, V. Dupont, and B. Le Bot, *Occurrence and toxicity of antimicrobial triclosan and by-products in the environment.* Environmental Science and Pollution Research, 2012. **19**(4): p. 1044-1065.
306. Jirasripongpun, K., Wongarethornkul, T. and Mulliganavin, S., *Risk Assessment of Triclosan Using Animal Cell Lines.* Kasetsart Journal Natural Science, 2008. **42**: p. 353-359.
307. Binelli, A., D. Cogni, M. Parolini, C. Riva, and A. Provini, *Cytotoxic and genotoxic effects of in vitro exposure to Triclosan and Trimethoprim on zebra mussel (Dreissena polymorpha) hemocytes.* Comparative Biochemistry and Physiology C-Toxicology & Pharmacology, 2009. **150**(1): p. 50-56.
308. Ciniglia, C., C. Cascone, R. Lo Giudice, G. Pinto, and A. Pollio, *Application of methods for assessing the geno- and cytotoxicity of Triclosan to C-ehrenbergii.* Journal of Hazardous Materials, 2005. **122**(3): p. 227-232.

References

309. Zuckerbraun, H.L., H. Babich, R. May, and M.C. Sinensky, *Triclosan: cytotoxicity, mode of action, and induction of apoptosis in human gingival cells in vitro*. European Journal of Oral Sciences, 1998. **106**(2): p. 628-636.
310. Farre, M., D. Asperger, L. Kantiani, S. Gonzalez, M. Petrovic, and D. Barcelo, *Assessment of the acute toxicity of triclosan and methyl triclosan in wastewater based on the bioluminescence inhibition of Vibrio fischeri*. Analytical and Bioanalytical Chemistry, 2008. **390**(8): p. 1999-2007.
311. Lindstrom, A., I.J. Buerge, T. Poiger, P.A. Bergqvist, M.D. Muller, and H.R. Buser, *Occurrence and environmental behavior of the bactericide triclosan and its methyl derivative in surface waters and in wastewater*. Environmental Science & Technology, 2002. **36**(11): p. 2322-2329.
312. Orvos, D.R., D.J. Versteeg, J. Inauen, M. Capdevielle, A. Rothenstein, and V. Cunningham, *Aquatic toxicity of triclosan*. Environmental Toxicology and Chemistry, 2002. **21**(7): p. 1338-1349.
313. Wilson, B.A., V.H. Smith, F. Denoyelles, and C.K. Larive, *Effects of three pharmaceutical and personal care products on natural freshwater algal assemblages*. Environmental Science & Technology, 2003. **37**(9): p. 1713-1719.
314. Ishibashi, H., N. Matsumura, M. Hirano, M. Matsuoka, H. Shiratsuchi, Y. Ishibashi, Y. Takao, and K. Arizono, *Effects of triclosan on the early life stages and reproduction of medaka Oryzias latipes and induction of hepatic vitellogenin*. Aquatic Toxicology, 2004. **67**(2): p. 167-179.
315. EC, *SSCP (Scientific Committee on Consumer Products), Opinion on Triclosan*. SSCP/1192/08, 2009.
316. FDA, *Nomination Profile Triclosan - Supporting Information for Toxicological Evaluation by the National Toxicology Program*. U.S. Food & Drug Administration Department of Health and Human Services, 2008.
317. Fang, J.-L., R.L. Stingley, F.A. Beland, W. Harrouk, D.L. Lumpkins, and P. Howard, *Occurrence, Efficacy, Metabolism, and Toxicity of Triclosan*. Journal of Environmental Science and Health Part C-Environmental Carcinogenesis & Ecotoxicology Reviews, 2010. **28**(3): p. 147-171.
318. Chen, J., J. Jiang, F. Zhang, H. Yu, and J. Zhang, *Cytotoxic effects of environmentally relevant chlorophenols on L929 cells and their mechanisms*. Cell Biology and Toxicology, 2004. **20**(3): p. 183-196.
319. Jiang, J., J.N. Chen, H.X. Yu, F. Zhang, J.F. Zhang, and L.S. Wang, *Quantitative structure activity relationship and toxicity mechanisms of chlorophenols on cells in vitro*. Chinese Science Bulletin, 2004. **49**(6): p. 562-566.
320. Hilliard, C.A., M.J. Armstrong, C.I. Bradt, R.B. Hill, S.K. Greenwood, and S.M. Galloway, *Chromosome aberrations in vitro related to cytotoxicity of nonmutagenic chemicals and metabolic poisons*. Environmental and Molecular Mutagenesis, 1998. **31**(4): p. 316-326.
321. Ensley, H.E., J.T. Barber, M.A. Polito, and A.I. Oliver, *Toxicity and Metabolism of 2,4-Dichlorophenol by the Aquatic Angiosperm Lemna-gibba*. Environmental Toxicology and Chemistry, 1994. **13**(2): p. 325-331.
322. Burkhardt, M., S. Zuleeg, R. Vonbank, K. Bester, J. Carmeliet, M. Boller, and T. Wangler, *Leaching of Biocides from Facades under Natural Weather Conditions*. Environmental Science & Technology, 2012. **46**(10): p. 5497-5503.

References

323. Mohr, S., R. Berghahn, W. Mailahn, R. Schmiediche, M. Feibicke, and R. Schmidt, *Toxic and Accumulative Potential of the Antifouling Biocide and TBT Successor Irgarol on Freshwater Macrophytes: A Pond Mesocosm Study.* Environmental Science & Technology, 2009. **43**(17): p. 6838-6843.
324. Mezcua, M., M.D. Hernando, L. Piedra, A. Aguera, and A.R. Fernandez-Alba, *Chromatography-mass spectrometry and toxicity evaluation of selected contaminants in seawater.* Chromatographia, 2002. **56**(3-4): p. 199-206.
325. Okamura, H., I. Aoyama, D. Liu, R.J. Maguire, G.J. Pacepavicius, and Y.L. Lau, *Fate and ecotoxicity of the new antifouling compound Irgarol 1051 in the aquatic environment.* Water Research, 2000. **34**(14): p. 3523-3530.
326. Zhou, X., H. Okamura, and S. Nagata, *Applicability of luminescent assay using fresh cells of Vibrio fischeri for toxicity evaluation.* Journal of Health Science, 2006. **52**(6): p. 811-816.
327. Okamura, H., T. Nishida, Y. Ono, and W.J. Shim, *Phytotoxic effects of antifouling compounds on nontarget plant species.* Bulletin of Environmental Contamination and Toxicology, 2003. **71**(5): p. 881-886.
328. Noguerol, T.N., S. Boronat, M. Casado, D. Raldua, D. Barcelo, and B. Pina, *Evaluating the interactions of vertebrate receptors with persistent pollutants and antifouling pesticides using recombinant yeast assays.* Analytical and Bioanalytical Chemistry, 2006. **385**(6): p. 1012-1019.
329. Liu, D., R.J. Maguire, Y.L. Lau, G.J. Pacepavicius, H. Okamura, and I. Aoyama, *Transformation of the new antifouling compound Irgarol 1051 by Phanerochaete chrysosporium.* Water Research, 1997. **31**(9): p. 2363-2369.
330. Liu, D., G.J. Pacepavicius, R.J. Maguire, Y.L. Lau, H. Okamura, and I. Aoyama, *Mercuric chloride-catalyzed hydrolysis of the new antifouling compound Irgarol 1051.* Water Research, 1999. **33**(1): p. 155-163.
331. Okamura, H., *Photodegradation of the antifouling compounds Irgarol 1051 and Diuron released from a commercial antifouling paint.* Chemosphere, 2002. **48**(1): p. 43-50.
332. Arufe, M.I., J. Arellano, M.J. Moreno, and C. Sarasquete, *Toxicity of a commercial herbicide containing terbutryn and triasulfuron to seabream (Sparus aurata L.) larvae: a comparison with the Microtox test.* Ecotoxicology and Environmental Safety, 2004. **59**(2): p. 209-216.
333. Gaggi, C., G. Sbrilli, A.M.H. Elnaby, M. Bucci, M. Duccini, and E. Bacci, *Toxicity and Hazard Ranking of S-Triazine Herbicides Using Microtox®), 2 Green Algal Species and a Marine Crustacean.* Environmental Toxicology and Chemistry, 1995. **14**(6): p. 1065-1069.
334. Villarini, M., G. Scassellati-Sforzolini, M. Moretti, and R. Pasquini, *In vitro genotoxicity of terbutryn evaluated by the alkaline single-cell microgel-electrophoresis "comet" assay.* Cell Biology and Toxicology, 2000. **16**(5): p. 285-292.
335. Plhalova, L., S. Macova, P. Dolezelova, P. Marsalek, Z. Svobodova, V. Pistekova, I. Bedanova, E. Voslarova, and H. Modra, *Comparison of Terbutryn Acute Toxicity to Danio rerio and Poecilia reticulata.* Acta Veterinaria Brno, 2010. **79**(4): p. 593-598.
336. Velisek, J., E. Sudova, J. Machova, and Z. Svobodova, *Effects of sub-chronic exposure to terbutryn in common carp (Cyprinus carpio L.).* Ecotoxicology and Environmental Safety, 2010. **73**(3): p. 384-390.
337. Velisek, J., A. Stara, J. Machova, P. Dvorak, E. Zuskova, M. Prokes, and Z. Svobodova, *Effect of terbutryn at environmental concentrations on early life*

References

stages of common carp (Cyprinus carpio L.). Pesticide Biochemistry and Physiology, 2012. **102**(1): p. 102-108.

338. Brix, R., N. Bahi, M.J.L. de Alda, M. Farre, J.-M. Fernandez, and D. Barcelo, *Identification of disinfection by-products of selected triazines in drinking water by LC-Q-ToF-MS/MS and evaluation of their toxicity*. Journal of Mass Spectrometry, 2009. **44**(3): p. 330-337.

339. Umweltbundesamt, *Fact Sheet Polymoschusverbindungen*.

340. Balk, F. and R.A. Ford, *Environmental risk assessment for the polycyclic musks, AHTN and HHCB - II. Effect assessment and risk characterisation*. Toxicology Letters, 1999. **111**(1-2): p. 81-94.

341. Api, A.M. and R.H.C. San, *Genotoxicity tests with 6-acetyl-1,1,2,4,4,7-hexamethyltetraline and 1,3,4,6,7,8-hexahydro-4,6,6,7,8,8-hexamethylcyclopenta-gamma-2-benzopyra n*. Mutation Research-Genetic Toxicology and Environmental Mutagenesis, 1999. **446**(1): p. 67-81.

342. Carlsson, G. and L. Norrgren, *Synthetic musk toxicity to early life stages of zebrafish (Danio rerio)*. Archives of Environmental Contamination and Toxicology, 2004. **46**(1): p. 102-105.

343. Randelli, E., V. Rossini, I. Corsi, S. Focardi, A.M. Fausto, F. Buonocore, and G. Scapigliati, *Effects of the polycyclic ketone tonalide (AHTN) on some cell viability parameters and transcription of P450 and immunoregulatory genes in rainbow trout RTG-2 cells*. Toxicology in Vitro, 2011. **25**(8): p. 1596-1602.

344. Seinen, W., J.C. Lemmen, R.H.H. Pieters, E.M.J. Verbruggen, and B. van der Burg, *AHTN and HHCB show weak estrogenic - but no uterotrophic activity*. Toxicology Letters, 1999. **111**(1-2): p. 161-168.

345. Schreurs, R., E. Sonneveld, J.H.J. Jansen, W. Seinen, and B. van der Burg, *Interaction of polycyclic musks and UV filters with the estrogen receptor (ER), androgen receptor (AR), and progesterone receptor (PR) in reporter gene bioassays*. Toxicological Sciences, 2005. **83**(2): p. 264-272.

346. Simmons, D.B.D., V.L. Marlatt, V.L. Trudeau, J.P. Sherry, and C.D. Metcalfe, *Interaction of Galaxolide (R) with the human and trout estrogen receptor-alpha*. Science of the Total Environment, 2010. **408**(24): p. 6158-6164.

347. van der Burg, B., R. Schreurs, S. van der Linden, W. Seinen, A. Brouwer, and E. Sonneveld, *Endocrine effects of polycyclic musks: do we smell a rat?* International Journal of Andrology, 2008. **31**(2): p. 188-193.

348. Schreurs, R.H.M.M., J. Legler, E. Artola-Garicano, T.L. Sinnige, P.H. Lanser, W. Seinen, and B. van der Burg, *In Vitro and in Vivo Antiestrogenic Effects of Polycyclic Musks in Zebrafish*. Environmental Science & Technology, 2003. **38**(4): p. 997-1002.

349. Witorsch, R.J. and J.A. Thomas, *Personal care products and endocrine disruption: A critical review of the literature*. Critical Reviews in Toxicology, 2010. **40**: p. 1-30.

350. Schreurs, R., E. Sonneveld, P.T. van der Saag, B. van der Burg, and W. Seinen, *Examination of the in vitro (anti) estrogenic, (anti) androgenic and (anti)dioxin-like activities of tetralin, indane and isochroman derivatives using receptor-specific bioassays*. Toxicology Letters, 2005. **156**(2): p. 261-275.

351. Schreurs, R., M.E. Quaedackers, W. Seinen, and B. van der Burg, *Transcriptional activation of estrogen receptor ER alpha and ER beta by polycyclic musks is cell type dependent*. Toxicology and Applied Pharmacology, 2002. **183**(1): p. 1-9.

References

352. Balk, F. and R.A. Ford, *Environmental risk assessment for the polycyclic musks AHTN and HHCB in the EU - I. Fate and exposure assessment.* Toxicology Letters, 1999. **111**(1-2): p. 57-79.
353. ECHA, *SVHC Support Document.* http://echa.europa.eu/documents/10162/d0f5c171-5086-49c3-a6a3-3a31cb4e08eb, 2009.
354. Follmann, W. and J. Wober, *Investigation of cytotoxic, genotoxic, mutagenic, and estrogenic effects of the flame retardants tris-(2-chloroethyl)-phosphate (TCEP) and tris-(2-chloropropyl)-phosphate (TCPP) in vitro.* Toxicology Letters, 2006. **161**(2): p. 124-134.
355. Haworth, S., T. Lawlor, K. Mortelmans, W. Speck, and E. Zeiger, *Salmonella mutagenicity test results for 250 chemicals.* Environmental mutagenesis, 1983. **5 Suppl 1**: p. 1-142.
356. Zeiger, E., B. Anderson, S. Haworth, T. Lawlor, and K. Mortelmans, *Salmonella Mutagenicity Tests. 5. Results from the Testing of 311 Chemicals.* Environmental and Molecular Mutagenesis, 1992. **19**: p. 2-141.
357. Sala, M., Z.G. Gu, G. Moens, and I. Chouroulinkov, *Invivo and Invitro Biological Effects of the Flame Retardants Tris(2,3-Dibromopropyl)-phosphate and Tris(2-Chloroethyl)-orthophosphate.* European Journal of Cancer & Clinical Oncology, 1982. **18**(12): p. 1337-1344.
358. Galloway, S.M., M.J. Armstrong, C. Reuben, S. Colman, B. Brown, C. Cannon, A.D. Bloom, F. Nakamura, M. Ahmed, S. Duk, J. Rimpo, B.H. Margolin, M.A. Resnick, B. Anderson, and E. Zeiger, *Chromosome-Aberrations and Sister Chromatid Exchanges in Chinese-Hamster Ovary Cells - Evaluations of 108 Chemicals.* Environmental and Molecular Mutagenesis, 1987. **10**: p. 1-175.
359. Ren, X., Y.J. Lee, H.J. Han, and I.S. Kim, *Cellular effect evaluation of micropollutants using transporter functions of renal proximal tubule cells.* Chemosphere, 2009. **77**(7): p. 968-974.
360. Ren, X., Y.J. Lee, H.J. Han, and I.S. Kim, *Effect of tris-(2-chloroethyl)-phosphate (TCEP) at environmental concentration on the levels of cell cycle regulatory protein expression in primary cultured rabbit renal proximal tubule cells.* Chemosphere, 2008. **74**(1): p. 84-88.
361. Liu, X., K. Ji, and K. Choi, *Endocrine disruption potentials of organophosphate flame retardants and related mechanisms in H295R and MVLN cell lines and in zebrafish.* Aquatic Toxicology, 2012. **114**: p. 173-181.
362. Higley, E., S. Grund, P.D. Jones, T. Schulze, T.-B. Seiler, U.L.-v. Varel, W. Brack, J. Wölz, H. Zielke, J.P. Giesy, H. Hollert, and M. Hecker, *Endocrine disrupting, mutagenic, and teratogenic effects of upper Danube River sediments using effect-directed analysis.* Environmental Toxicology and Chemistry, 2012. **31**(5): p. 1053-1062.
363. Flaskos, J., W.G. McLean, and A.J. Hargreaves, *The Toxicity of Organophosphate Compounds Towards Cultured PC12 Cells.* Toxicology Letters, 1994. **70**(1): p. 71-76.

8. Annex

8.1. List of Chemicals

Product	Company
CHO Cells	ECACC; Salisbury, UK
T47D Cells	BioDetectionSystems (BDS); Amsterdam, NL
Salmonella typhimurium TA98	Xenometrix; Allschwil, CH
Salmonella typhimurium TA100	Xenometrix; Allschwil, CH
17ß-Ethinylestradiol	BDS; Amsterdam, NL
2-Aminoanthracene	Xenometrix; Allschwil, CH
2-Nitrofluorene	Xenometrix; Allschwil, CH
4-Nitroquinolone-N-Oxide	Xenometrix; Allschwil, CH
Acetic acid	AppliChem; Darmstadt, D
Ames MPF Aqua 98/100 Test Kit (J10-210)	Xenometrix; Allschwil, CH
Ames Exposure Medium Solution A (10x)	Xenometrix; Allschwil, CH
Ames Exposure Medium Solution B (10x)	Xenometrix; Allschwil, CH
Ampicillin	Xenometrix; Allschwil, CH
Ampuwa (sterile water)	Fresenius Kabi; Bad Homburg, D
Blood agar plates	Oxoid; Wesel, D
CO_2 (N45)	Air Liquide; Düsseldorf, D
Dimethylsulfoxide (DMSO)	Sigma-Aldrich; Steinheim, D
DMEM F12 with Phenolred	Gibco; Karlsruhe, D
DMEM F12 without Phenolred	Gibco; Karlsruhe, D
Ethylenediaminetetraacetic acid (EDTA)	Sigma-Aldrich; Steinheim, D
N-ethyl-N-nitrosourea (ENU)	Sigma-Aldrich; Steinheim, D
Fetal Calf Serum (FCS)	Gibco; Karlsruhe, D
Gentamycin	c.c.pro; Oberdorla, D
GlowMix	BDS; Amsterdam, NL
Growth Medium (Ames)	Xenometrix; Allschwil, CH
HAM's F12	c.c.pro; Oberdorla, D
Hydrochloric acid (HCl)	Merck; Darmstadt, D

Annex

Product	Company
Indicator Medium (Ames)	Xenometrix; Allschwil, CH
LDH I (Reconstitution Solution)	Xenometrix; Allschwil, CH
LDH II (NADH)	Xenometrix; Allschwil, CH
LDH III (Pyruvate)	Xenometrix; Allschwil, CH
L-Glutamine	c.c.pro; Oberdorla, D
Liquid nitrogen	Air products; Bochum, D
Low melting point (L.M.P.) Agarose	Invitrogen; Paisley, UK
Lysis solution ER Calux	BDS; Amsterdam, NL
MTT	Sigma-Aldrich; Steinheim, D
MMAIII (Monomethylarsonous acid)	Argus Chemicals; Vernio, I
Sodiumchloride (NaCl)	Merck; Darmstadt, D
Sodiumhydroxide (NaOH)	Sigma-Aldrich; Steinheim, D
N-Laurylsarcosine Sodium Salt	Sigma-Aldrich; Steinheim, D
Non-essential amino acids (NEAA)	c.c.pro; Oberdorla, D
NR I (Washing solution)	Xenometrix; Allschwil, CH
NR II (Labeling solution)	Xenometrix; Allschwil, CH
NR III (Fixing solution)	Xenometrix; Allschwil, CH
NR IV (Solubilization solution)	Xenometrix; Allschwil, CH
PAN I Cytotoxicity Kit (PAN I 96.1200)	Xenometrix; Allschwil, CH
Phosphate Buffered Saline (PBS)	Gibco; Karlsruhe, D
Reference water	BDS; Amsterdam, NL
S9 Fraction	Xenometrix; Allschwil, CH
S9 Buffer Salts	Xenometrix; Allschwil, CH
S9 G-6-P	Xenometrix; Allschwil, CH
S9 NADP	Xenometrix; Allschwil, CH
Sodium dodecyl sulfate (SDS)	Sigma-Aldrich; Steinheim, D
SRB I (Washing solution)	Xenometrix; Allschwil, CH
SRB II (Fixing solution)	Xenometrix; Allschwil, CH
SRB III (labeling solution)	Xenometrix; Allschwil, CH
SRB IV (Rinsing solution)	Xenometrix; Allschwil, CH

Annex

Product	Company
SRB V (Solubilization solution)	Xenometrix; Allschwil, CH
Stripped FCS (Fetal Calf Serum)	BDS; Amsterdam, NL
SYBR-Green®	Sigma-Aldrich, Steinheim, D
Triton-X	Merck; Darmstadt, D
Trizma	Sigma-Aldrich; Steinheim, D
Trypsin-EDTA	c.c.pro; Oberdorla, D
XTT I (Substrate)	Xenometrix; Allschwil, CH
XTT II (Buffer)	Xenometrix; Allschwil, CH

Annex

8.2. List of Materials

Material	Catalogue number	Company
Cell culture flask (25 cm^2)	90025	TPP; Trasadingen, CH
Cell culture flask (75 cm^2)	90076	TPP; Trasadingen, CH
Cell culture flask (150 cm^2)	90151	TPP; Trasadingen, CH
Chamber Slides (8 chambers)	354118	BD Falcon; Heidelberg, D
Cover glasses	24 x 60 mm	Engelbrecht; Edermünde, D
Cuvettes	12.5x12.5x45 mm	Brand; Wertheim, D
Erlenmeyer flask (100 mL)	21216240	Schott; Mainz, D
Gelbond Film	53734	Lonza; Basel, CH
Microscope slides	631-1554	VWR; Darmstadt, D
24-well tissue culture plate	662160	Greiner Bio One; Frickenhausen, D
96-well microtiter plate (transparent)	167008	Nalge Nunc International; Wiesbaden, D
96-well microtiter plate (white)	655075	Greiner Bio One; Frickenhausen, D
96-well microtiter plate (transparent)	353072	BD Falcon; Heidelberg, D
384-Well microtiter plate	164688	Nalge Nunc International; Wiesbaden, D
Minisart® Filter (0.2 µm)	16534	Sartorius Stedim Biotech; Göttingen, D
Microcentrifuge tubes (1.5 mL)	780400	Brand; Wertheim, D
QUANTOFIX® Peroxide 25	91319	Macherey-Nagel; Düren, D

Annex

8.3. List of Equipment

Equipment	Name	Company
8-channel Pipette	Finnpipette®	Thermo Fisher Scientific; Bonn, D
8-channel pipette (electronic)	07 CH81250	Süd-Laborbedarf; Gauting, D
Autoclave	Dampfsterilisator 25-T 120°C	Thermo Fisher Scientific; Bonn, D
Cold chamber (4 °C)	Integra P 2.5 – 230 S	Viessmann; Allendorf, D
Comet Assay Software	Comet 4	Perceptive Instruments; Suffolk, UK
CO_2-Incubator	Model 3548 S/N	Heraeus; Hanau, D
Electrophoresis Chamber	Classic CSSU911	Thermo Fisher Scientific; Bonn, D
Electrophoresis Powersupply	EC 105	E-C Apparatus Corporation; St. Petersburg, US
Fluorescence microscope	Leica DMBL	Leica DMBL; Leitz GmbH & Co KG, Wetzlar, D
Fridge (4 °C)	Premium	Liebherr; Ochsenhausen, D
Fridge (4 °C) - Freezer (-20 °C)	Privileg 064.476 5	München, D
Ice machine		INCO Ziegra; Isernhagen, D
Laminar Flow work bench II	Hera Safe, Typ HS 18	Thermo Electron Corporation; Langenselbond, D
Laminar Flow work bench I	Cytostatic safety cabinet H-130	Berner International GmbH; Elmshorn, D
Light microscope	Labovert FS	Leitz GmbH & Co. KG; Wetzlar, D
Magnetic stirrer	IKAMAG®RET	Janke und Kunkel GmbH & Co. KG; Staufen i. Br, D
Microplate reader (Absorbance at 340, 480, 540 and 595 nm; Luminescence)	TECAN GENios	TECAN Trading AG; Männedorf, Ch
Microscope camera	Marlin - F046B	Allied Vision Technologies GmbH; Stadtroda, D
Multipette® Plus	3122 000.035, 3122 000.051	Eppendorf AG; Hamburg, D
Neubauer chamber	Neubauer Improved	Brand GmbH; Wertheim, D
Liquid nitrogen tank	MVE xc 34/18	Cryodepot/nexAir, LLC; Memphis, US
ph Meter	pH 521	Wissenschaftlich-technische Werkstätten; Weilheim i. OB, D
Pipetting aid	Pipetus	Hirschmann; Eberstadt/Württ, D
Serological pipettes	1 – 50 mL	Greiner Bio-One; Frickenhausen, D
Shaker	IKA-Vibrax-VXR	Janke und Kunkel GmbH & Co. KG; Staufen i. Br, D
Storage boxes		
Vortex	REAX 2000	Heidolph Instruments GmbH & Co. KG; Schwalbach, D

Annex

Equipment	Name	Company
Weigh	LE 225D-0CE	Sartorius AG; Göttingen, D
Water bath (37 °C)	GFL Typ 1002	GFL Gesellschaft für Labortechnik GmbH; Burgwedel, D
Water bath (78 °C)	GFL Typ 1004	GFL Gesellschaft für Labortechnik GmbH; Burgwedel, D
µL Pipettes with tips	3120 000.020, 3120 000.046, 3120 000.054, 3120 000.062	Eppendorf AG; Hamburg, D

Acknowledgements

At this point I would like to thank all those people who participated in the successful completion of my PhD thesis:

- Prof. Dr. Elke Dopp for the opportunity to pursue a PhD and for being a supportive mentor and for her encouragement throughout this time.

- Prof. Dr. Alfred V. Hirner for his support as a reviewer.

- Melanie Gerhards and Ricarda Zdrenka for being not only great colleagues but also friends and their helpful discussions and support during this time.

- Dr. Jochen Türk, Andrea Börgers and Claudia vom Eyser for preparing and providing the water samples, performing the oxidation experiments and also for their helpful discussions and support and for making it a successful project.

- Dr. Kai Bester as well as his working group for also providing water samples and performing the oxidative treatment.

- My colleagues from the Institute of Hygiene and Occupational Medicine at the University Hospital Essen especially Gabriele Zimmer and Ute Zimmermann.

- All my colleagues at the IWW for the opportunity to finish my thesis, for their support and for making it a great place to work.

- Dr. Peter Behnisch for his advice and helpful discussions as well his colleagues from BDS for their support and troubleshooting with the ER Calux.

- Dr. Tamara Grummt from the Umweltbundesamt for her helpful discussions and suggestions regarding this project.

I am also grateful to my family and friends for their support, encouragement and patience throughout this time.

i want morebooks!

Buy your books fast and straightforward online - at one of world's fastest growing online book stores! Environmentally sound due to Print-on-Demand technologies.

Buy your books online at
www.get-morebooks.com

Kaufen Sie Ihre Bücher schnell und unkompliziert online – auf einer der am schnellsten wachsenden Buchhandelsplattformen weltweit! Dank Print-On-Demand umwelt- und ressourcenschonend produziert.

Bücher schneller online kaufen
www.morebooks.de

VDM Verlagsservicegesellschaft mbH
Heinrich-Böcking-Str. 6-8   Telefon: +49 681 3720 174   info@vdm-vsg.de
D - 66121 Saarbrücken   Telefax: +49 681 3720 1749   www.vdm-vsg.de

Printed by Books on Demand GmbH, Norderstedt / Germany